The U.S. Cybersecurity and Intelligence Analysis Challenges

John Michael Weaver

The U.S. Cybersecurity and Intelligence Analysis Challenges

palgrave
macmillan

John Michael Weaver
Department of History and Political
Science
York College of Pennsylvania
York, PA, USA

ISBN 978-3-030-95840-4   ISBN 978-3-030-95841-1   (eBook)
https://doi.org/10.1007/978-3-030-95841-1

Cover illustration: © Melisa Hasan

This Palgrave Macmillan imprint is published by the registered company Springer Nature
Switzerland AG
The registered company address is: Gewerbestrasse 11, 6330 Cham, Switzerland

PREFACE

The forthcoming analysis in this book looked at taking a conventional understanding of the four instruments of national power (diplomacy, information, military, and economic measures/D.I.M.E.) and turning the tables to see how potential adversarial powers could use these against the national security interests of the United States. Moreover, it focused on qualitative research regarding the cyber threat that has continually beleaguered this nation by malevolent actors mostly over the last seven years. The study also affords consideration to how potential adversarial non-state actors and nation-states can implement the instruments of national power through the application of a new model named the York Intelligence Red Team Model-Cyber (Modified) [YIRTM-C (M)] using sources guided by the Federal Qualitative Secondary Data Case Study Triangulation Model to arrive at results.

John Michael Weaver
Department of History
and Political Science
York College of Pennsylvania
York, PA, USA

CONTENTS

About the Author

John Michael Weaver is an Associate Professor of Intelligence Analysis at York College of Pennsylvania (USA), a retired DOD civilian from the United States' Intelligence Community and has served as an officer in the U.S. Army (retiring at the rank of lieutenant colonel). He has lived and worked on four continents and in 19 countries spending nearly eight years overseas (on behalf of the US government). His experience includes multiple combat deployments, peace enforcement, peacekeeping, humanitarian relief, and disaster assistance support in both conventional and unconventional/non-traditional units. John has trained and certified multinational NATO reconnaissance teams based in The Netherlands, Germany, and Spain for worldwide deployment in full-spectrum mission sets. He has also personally led several reconnaissance missions throughout Europe, the Middle East, and Asia (including multiple missions in Afghanistan). He has received formal training/certification in the following areas from the US Department of Defense: Survival/Evasion/Resistance/Escape (high risk), communications equipment & communications planning (FM radio, landline & satellite communications, encryption, and the use of cryptographic devices), digital camera use & digital photography courses, US Joint Forces Command joint intelligence course, US Special Operations Command counterintelligence awareness course (USSOCOM CI), US Joint Forces Command counterintelligence awareness training (USJFCOM CI), counterinsurgency course, joint antiterrorism course, defense against suicide bombing course, dynamics of international terrorism, homeland security and defense course, the joint special operations task force course (JSOTF), defensive driving course, vehicle emergency drills (battle drills), composite risk management, the airborne and air assault schools, and more. Additionally, he graduated from NATO's Combined Joint Operations Center course in Oberammergau Germany, the Air Command and Staff College, and the Joint & Combined Warfighting School. John earned a Bachelor of Arts degree in business management from Towson University in 1990, graduated from Central Michigan University with a Master of Science in Administration degree in 1995, earned a Master of Operational Arts and Science degree from the U.S. Air Force's Air University in 2004, and graduated from the University of Baltimore with a Doctorate in Public Administration in 2013.

LIST OF FIGURES

CHAPTER 1

Introduction and Background

Abstract The use of the cyber domain for conflict is a relatively recent phenomenon. This chapter introduces the reader to cybersecurity threats and provides background information.

Keyword CNA · CNE · CNO · Cyber

This work looks to implement what Boyer calls the scholarship of integration (Glassick 2000). It does so by considering a series of micro-case studies looking at four nation-states and two non-state actors. One of the prevailing issues of contemporary times centers on the issue of cyber threats confronting western nations (Weaver 2017). When focusing attention to the topic of information and cyber-warfare, information is often seen as far more advantageous than money and is even more valued than currency because it is through information that one can attain more wealth (Bruce et al. 2004, 11). It can influence opinions, shape actions, and through using what is now termed "deep fakes" can modify the actual presentation of speeches by leaders to convey anything other than what they actually stated (with extreme realism).

These threats are increasingly more disruptive and destructive, and most nations' infrastructure is extremely vulnerable to them (GAO-16-332 2016). As individuals, companies, and governments at all levels rely

J. M. Weaver, *The U.S. Cybersecurity and Intelligence Analysis,*
https://doi.org/10.1007/978-3-030-95841-1_1

1

more on information technology, collaborative tools like Google Drive (Google Docs, Google Sheets, and the view/edit function), and Kubernetes, information could become vulnerable to exploitation, and as a result, a lot of inherent risks exists in today's world. Compounded with the use of asymmetric and hybrid warfare, these threats are real.

The Cyber Domain

The use of the cyber domain for conflict is a relatively recent phenomenon. Indeed, debates are ongoing as to whether cyber should be regarded as a fifth-domain analogous to the physical military theaters of air, land, sea, and space and whether there has yet been any real incident akin to conflict on those domains meeting the threshold of what constitutes warfare (McGuffin and Mitchell 2014). The burgeoning of cyber is marked by several incidents and prolific cyber-weapons: the Central Intelligence Agency's (CIA's) use of the 'logic bomb' is often considered the first case of cyber operations concerning national security; these also include the North Atlantic Treaty Organization's (NATO's) PROMIS; the emergence of non-state hackers; the Russian's 'Moonlight Blaze' (which gained information about American missile targeting systems); and the Chinese 'Titan Rain,' just to name but a few (Lakomy 2013, 108). A critical event concerning the militarization of cyber within classic geopolitics is Russia's actions in Estonia, which is regarded by some as the first 'cyber war' "since computer networks were used to paralyze the critical infrastructure of a nation-state" (Lakomy 2013, 106). Lastly, the famous 'Stuxnet' virus is routinely referenced as the beginning of a "new era of cyber-warfare and suggested that this new type of cyber-weapon had a similar meaning for international security as the bombing of Hiroshima and Nagasaki" (Lakomy 2013, 106). This statement underscored that cyber, like nuclear weapons over time, has changed the actual structural conditions of conflict and warfare while it increased the 'grey zone' range of operations once reserved for classical espionage. An empirical question regarding cyber is to what degree will cyber continue to be used, and within what capacities, in terms of offensive attacks (Weaver and Johnson 2020). While most concerns for cyber defense examine the integration of civilian infrastructure with the internet of things (IoT), offensive cyber tactics, techniques, and procedures (TTPs) may be used to attack a military's increasing dependence on networked technology, even that which is offline. Cyber may also play a part in hybrid warfare, where attackers may "leverage this expanded attack surface through target sets that generate effects in both the information and physical domains" through direct or *nth* level multiplier effects (Gendron 2013; Leuprecht et al. 2019, 389). These actors could also help enhance their anonymity using virtual private network (VPN) technology, The Onion Router (TOR) browser (and surfing the 'dark web'), etc., making it very difficult to trace the origin of cyber events.

Thus, cyber operations can be utilized to undermine the integrity of a country by hindering the state's ability to pursue its interests (however defined) through immobilization, which may also see the manifestation of physical destruction in the classic sense of warfare (Weaver and Johnson 2020). Further, cyber operations could help undermine the stability of a society that is reminiscent of classic disinformation campaigns. This might be true as the diversification of actors and their access to technology, combined with decreases of violent geopolitical conflict between major state powers (Kshetri 2013), suggests that digital and cyber-based threats will increase in sophistication and prevalence as they are used to elicit specific political, social, and economic outcomes (Weaver and Johnson 2020).

Politically, Lakomy essentially argues that "Cyber-warfare challenges the security policies of all industrial states and the lack of a clear international mechanism to coordinate responses has increased the need for independent action to be undertaken in a spinoff of an arms race. In short, each state is forced to develop its plan of action with or without its allies" (2013, 108). Thus, cyber dangers may prioritize national action and self-interest from a classical focus on *raison d'état* as the international order is too slow to adapt and other competitive pressures may emerge (e.g., trade wars) (Weaver and Johnson 2020).

The seeming lack of coordinated policy between allied states on cyber-security may partially be linked to this structural shift in addition to the seeming inability of international mechanisms, including organizations, regimes, and laws, to adapt to what is seen as the new structural reality (Weaver and Johnson 2020). This inability for the liberal international order to adapt to rising challenges, and by extension, the difficulties encountered by individual states, is partially due to the rapid acceleration of technological change. As Adams argues, "owing to market incentives, innovation in functionality is outpacing innovation in security" (2016, 1). However, the lack of coordinated international policy on cybersecurity may also be the result of the discursive dominance of cyber-'war' rather than, say, cyber-'crime' (Levin and Goodrick 2013). The former often is indicative of competition among nation-states, firmly entrenching cybersecurity in the realm of national defense whereas the latter suggests the need to combat illegitimate criminal organizations and enterprises that may hold relevance to national security but also is inclusive of the cooperation of non-military actors, such as policing forces across various boundaries, both state and non-state. Overall, the amalgam of this structural shift inherent to the wide array of cyber technologies, and the rhetorical focus on warfare and military theaters has led many nations to pursue a cybersecurity strategy predicated on 'resilience' underpinned by public–private partnerships (P2Ps) to induce industry to innovate and take up part of the defense burden given that a great deal of cyber is privately owned (e.g., CADSI 2019), creating a "polycentric governance architecture" (Dupont 2018, 26).

However, Carr (2016) argues that there is an inherent challenge in this strategy, mostly due to contradictions between the goals (of private capital) and the needs (of national security), or more broadly, the difference between private interests and public goods (Weaver and Johnson 2020). Carr goes on to argue that a central problem to P2Ps is "the expectation that the private sector will invest in cybersecurity beyond its cost/benefit analysis to fully accommodate the public interest - in other words, to ensure national security" (2016, 60). Likewise, Grayson and O'Higgins (2018, 30) make the point that governments need to play a central role in cybersecurity efforts through the incorporation of private actors rather than relying on them. However, even when governments are accorded a central role in defending their populations against cyber threats, this may create a 'wicked problem' for liberal democracies such as that of the United States. The 'wicked problem' refers to the contradiction between enabling freedom and ensuring security in democratic societies, or "where to draw the line between adequate security, reasonable cost, and personal freedom" (Malone and Malone 2013, 158). The 'wicked problem' is especially problematic when considering the ideational aspect of cyber threats, i.e., operations designed to create disinformation among society, and which may be combined with other forms of attack (more recently termed 'hybrid warfare'). Cyber threats believed to be linked to disinformation campaigns are especially problematic due to the fact that they undermine and challenge the ontological security of societies (the taken-for-granted faith that the world is as it appears, which underlies our sense of trust in that world and its representative institutions) but protecting against these hybrid threats might also lead to the onset of ontological insecurity. There are potential contradictions for democratic governments seeking to protect themselves against hybrid threats. Hybrid threats, by the nature of their hybridity, are omnipresent and consequently may invite an equally totalizing response by state agencies and institutions. In turn, this presents a challenge for policymakers in liberal democratic societies that provide specific freedoms (e.g., speech, press, and peaceful dissent) while they also need to balance security concerns.

Further, more recent examples of disinformation campaigns point toward the complex organic and networked form of these threats when combined with digital technologies; this is inclusive of cyber operations (Weaver and Johnson 2020). These disinformation campaigns can assume a life of their own as they are morphed and reproduced by individuals in social networks (both digital and physical) (Weaver and Johnson 2020). Disinformation campaigns are often integrated with and spread by individuals in a variegated mix of social groups, thereby affecting and being affected by the underlying practices and organization of said groups. The policy challenge then speaks to the contradictory nature of issue of defending against disinformation while simultaneously protecting the freedom of expression and the integrity of information itself, which is indicative of the discursive focus by governments on their need for building 'resilience' (Mälksoo 2018).

One then can ask, what is the relevance and why is this important? Cyber-attacks are rising and being executed by individuals, nation-states, and non-state actors like terror organizations (Norris et al. 2019). This is particularly disconcerting as the number of networked devices in the average home grew from two just a few years ago to 15 today (Norris et al. 2019). Moreover, at the end of 2019, estimates of world attacks amounted to roughly $2.1 trillion (Juniper Research 2015).

Cyber-Strategies of the Nation

United States

The National Security Strategy (NSS) of the United States serves as one of the most important unclassified writings in the world today concerning national security. The strategy provides clear direction and guidance to those who work on federal-related issues for the government of the

United States and its allies. It also succinctly articulates to both allies and potential belligerents where security efforts should focus. Accordingly, this strategic guidance is adopted by the Departments of State, Defense, and Homeland Security to look at a top to bottom approach to request resources, assign missions, and allocate manpower to work toward desired ends. The NSS is also used by the Intelligence Community in conjunction with the *national intelligence priority framework* and other intelligence products produced by allies to help identify requirements and the subsequent tasking of intelligence collection platforms.

Readers will quickly discern the linkages of key issues and topics that are vulnerable to cyber exploitation and attacks through an analysis of the NSS. Moreover, both computer network attacks (CNA) and computer network exploitation (CNE) fall under the overarching concept of computer network operations (CNO) and are determined to be offensive in nature (Brantley 2016). These include among others, a nation's financial sector, the county's military, its science and technology base, the energy sector, the transportation infrastructure, the health industry, and more (NSS 2017). CNO efforts (expanded to include CNA and CNE) are often employed to disrupt, deny, deceive, degrade, and/or destroy various computer systems, their networks, and services (Rudner 2013, 454). The United States has underscored the relevance of cybersecurity to the continuance of prosperity of western nations to foster a more reliable, secure, interoperable, and open internet while mitigating cyber threats (NSS 2017). This is particularly relevant when looking at the linkage of the United States' economy through trade.

The Defense Department's Cyber Strategy expands on the NSS in greater detail concerning cyber-related issues. In April of 2015, Secretary of Defense Carter explained in his introductory letter that the United States relies extensively on the internet and cyberspace for a full menu of critical services. He went on to explain in his opening remarks that this dependency exposes individuals, a nation's military, businesses, schools, and the government to the vulnerabilities that could be realized through exploitation and attack (Cyber Strategy 2015). What's more is that this document attests to the importance of the D.I.M.E. to defend cyberinfrastructure from likely attackers (Cyber Strategy 2015, 2).

More recently, the U.S. Department of Defense (DOD) stated in the 2018 update to its cyber strategy that the DOD was significantly impacted by cybersecurity threats (Cyber Strategy 2018). The current objectives of the Department of Defense, according to the Cyber Strategy document, are to build a more lethal joint force, compete with and deter threats in cyberspace, strengthen alliances, and attract new partnerships that will hopefully reform the department and cultivate talent where needed (McDowell et al. 2019).

To build a more lethal Joint Force, the Defense Department will actively pursue the development of cyber capabilities for both warfighting and non-combat operations to counter malicious state and non-state actors while simultaneously maintaining flexibility to Joint Force commands to help them operate smoothly during day-to-day operations, which includes rapidly adapting to new threats and technologies in cyberspace (Cyber Strategy 2018). Part of building a more lethal Joint Force that can operate at the highest speeds across different networks includes being able to support analytics to identify malicious cyber activity that ensures tailored counter operations specific to that activity (Cyber Strategy 2018).

Competing (while also deterring) in cyberspace is a top priority of this department which would conclude that the United States more broadly would use all means available to deter malicious cyber activities and counter cyber campaigns to halt or intercept cyber threats (Cyber Strategy 2018). Working in close step with the private sector is also a major pillar in deterring cyber-attacks through the open sharing of information on these events that would allow the private sector to evolve its cybersecurity to defend infrastructures and networks (McDowell et al. 2019). The Defense Department's plan to enhance the strength of alliances and to attract new partnerships in the fight against cyber-crime will be brought about in several ways (McDowell et al. 2019). The first of these is centered on a strong focus to build trust with private sector organizations and forming partnerships with friendly nations. These partnerships are intended to help share and support cybersecurity activities because of how largely the private sector operates within the United States (Cyber Strategy 2018). Not only will these public–private partnerships benefit from private companies' expertise, but they will also help support norms of responsible behavior that will hopefully lead to the protection of cyberspace to defend infrastructure privately (Cyber Strategy 2018).

As a result of new threats directed at the United States, the Defense Department is undergoing a major reformation of its cyber operations (McDowell et al. 2019). This reformation is expected to more aptly

enable the Defense Department to adapt to its new cyberspace accountability, thereby holding public and private sector actors accountable for
their actions and for them to seek out solutions that are not just reliable,
but affordable as well (Cyber Strategy 2018).

To better prepare this nation for threats of cyber, President Biden
enacted a three-pronged approach (Biden Press Release 2021). It turns to
(1) modernizing the nation's defenses of federal, state, and local government infrastructure, (2) rebuilding the U.S. presence globally, and (3)
ensuring that the country is well postured to compete on the world stage
(Biden Press Release 2021). Ultimately, the U.S. wants to create a system
of metrics by which the government can measure effectiveness against
cybersecurity threats (Biden Press Release 2021). This in turn will be
coupled with Biden's Industrial Control System Cybersecurity Initiative
to better provide threat visibility, indicators, and detection, in addition to
warnings (Biden Press Release 2021).

DOD's final action of adaptation to this new cyber world order
includes cultivating talent within the department. This includes more
extensive investment in attracting and developing talent, and overall
building a better recruitment platform for an up-to-date cyber work-
force that is constantly ready to counter any cyber-attack at any time
(McDowell et al. 2019). Primarily, the DOD will seek to fill its ranks
with experienced military service members, civilian employees who are
cyber-savvy, and contracted service personnel, who have cybersecurity
skills to co-deter cyber events (Cyber Strategy 2018). As pointed out by
the DOD, deterrence is and will remain a major component of its defensive strategy by using early warning capabilities and fast responses to fend
off cyber-attackers (McDowell et al. 2019).

The United States' Intelligence Community (IC), in conjunction with
the DOD, stresses the threat posed by cyber (NIS 2019, 3). The former
Director of National Intelligence, Dan Coates, had made it clear that
cyber is an ever-evolving threat to national security. More pointedly, the
U.S. National Intelligence Strategy (NIS) points out that cyber threats
can undermine public confidence in governance and institutions while
weakening the economy (NIS 2019, 4). Importantly, the NIS prioritizes
cyber at a higher level of threat than terrorism and the proliferation of
weapons of mass destruction (WMD) because of the significant rise in
cyber incidents directed at the United States in recent years at a level
much greater than with terrorism and WMD (NSS 2017, 7).

In a speech given at college graduation back in 2018, Gina Haspel, who headed the U.S. Central Intelligence Agency, stated that cyber was one of the critical issues that kept her up at night (Haspel 2018). Cyber-weapons are seemingly unique and are qualitatively different from kinetic ones (Leuprecht et al. 2019, 382).

When turning to U.S. doctrine, one sees that it is advantageous operationally to collect, process, and disseminate information without interruption while also simultaneously exploiting an adversary's ability to do the same is of paramount concern (JP 3-13 2013). Those who are adept at implementing cyber threats are driven in part because that knowledge helps generate greater knowledge, and knowledge is often akin to power. Cyber-attacks often see not only the manifestation of physical damage to networks and hardware systems, but often realize intangible effects to a state, its infrastructure, and communities that are extremely difficult to quantify (Al-Ahmed 2013). In the cyber field, one can see the fruition of such occurrences as network degradation, destruction, and the alteration and manipulation of information (JP 3-12 2014).

What makes the cyber domain unique is that it is not confined by geographic space and can be near instantaneously linked to any part of the planet (Brecher 2012). Cyber, like information communications technology more generally, has fostered the compression of time and space such that the United States does not enjoy the relative safety from attacks borne out of physical distance. Moreover, cyber brings about a series of political challenges that confront governments domestically, bilaterally, and internationally; it raises concerns about attribution, deterrence, and response (Leuprecht et al. 2019, 382).

References

Adams, J. 2016. *Canada and Cyber*. Ontario, Canada: Canadian Global Affairs Institute.

Al-Ahmed, W. 2013. A Detailed Strategy for Managing Corporation Cyber War Security. *International Journal of Cyber-Security and Digital Forensics* 2 (4): 1–9.

Biden Press Release. 2021. Background Press Call on Improving Cybersecurity of U.S. Critical Infrastructure. https://www.whitehouse.gov/briefing-room/press-briefings/2021/07/28/background-press-call-on-improving-cybersecurity-of-u-s-critical-infrastructure/. Accessed 28 July 2021.

Brantley, Aaron F. 2016. Cyber Actions by State Actors: Motivation and Utility. *CBO of Intelligence and Counterintelligence* 27 (3): 465–484.

Brecher, A.P. 2012. Cyberattacks and the Covert Action Statute: Towards a Domestic Legal Framework for Offensive Cyberoperations. *Michigan Law Review* 111 (3): 423–452.

Bruce, C., S. Hicks, and J. Cooper (eds.). 2004. *Exploring Crime Analysis: Readings on Essential Skills*, 2nd ed. North Charleston, SC: BookSurge LLC.

CADSI. 2019. *From Bullets to Bytes: Industry's Role in Preparing Canada for the Future of Cyber Defence*.

Carr, Madeline. 2016. Public–Private Partnerships in National Cyber-Security Strategies. *International Affairs* 92 (1): 43–62.

Cyber Strategy. 2015. The Department of Defense Cyber Strategy. http://www.defense.gov/Portals/1/features/2015/0415_cyber-strategy/Final_2015_DoD_CYBER_STRATEGY_for_web.pdf. Accessed 16 Feb 2016.

Cyber Strategy. 2018. The Department of Defense Cyber Strategy. https://media.defense.gov/2018/Sep/18/2002041658/1/1/1/CYBER_STRATEGY_SUMMARY_FINAL.PDF. Accessed 23 Sept 2018.

Dupont, B. 2018. Mapping the International Governance of Cybercrime. In *Governing Cybersecurity in Canada, Australia and the United States*, ed. C. Leuprecht and S. MacLellan, 23–28. Centre for International Governance Innovation.

GAO-16-332. 2016. Civil Support DOD Needs to Clarify Its Roles and Responsibilities for Defense Support of Civil Authorities, During Cyber Incidents. http://www.gao.gov/assets/680/676322.pdf. Accessed 6 Apr 2016.

Gendron, A. 2013. Cyber Threats and Multiplier Effects: Canada at Risk. *Canadian Foreign Policy Journal* 19 (2): 178–198.

Glassick, Charles E. 2000. Boyer's Expanded Definitions of Scholarship, the Standards for Assessing Scholarship, and the Elusiveness of the Scholarship of Teaching. *Academic Medicine* 75 (9): 877–880.

Grayson, T., and O'Higgins, Brian. 2018. Cyber Scaffolding: Proposing a National Organization to Support the Canadian Economy and Public Safety. In *Governing Cybersecurity in Canada, Australia, and the United States*, ed. C. Leuprecht and S. MacLellan, 29–32. Centre for International Governance Innovation.

Haspel, Gina. 2018. Remarks for Central Intelligence Agency Director Gina Haspel–McConnell Center at the University of Louisville. https://www.cia.gov/news-information/speeches-testimony/2018-speeches-testimony/remarks-for-central-intelligence-agency-director-gina-haspel-mcconnell-center-at-the-university-of-louisville.html. Accessed 17 June 2019.

JP 3-12. 2014. Cyberspace Operations. http://edocs.nps.edu/2014/October/jp3_12R.pdf. Accessed 16 June 2019.

JP 3-13. 2013. Information Operations. file:///C:/Users/weave/AppData/Local/Microsoft/Windows/INetCache/IE/FLXPPAVZ/759867.pdf. Accessed 16 June 2019.

Juniper Research. 2015. Cybercrime Will Cost Businesses Over $2 Trillion by 2019. News Release, May 12. https://www.juniperresearch.com/press/pre sseleases/cybercrime-cost-businesses-over-2trillion. Accessed 20 Jan 2019.

Kshetri, N. 2013. Cybercrime and Cyber-Security Issues Associated with China: Some Economic and Institutional Considerations. *Electronic Commerce Research* 13 (1): 41–49.

Lakomy, M. 2013. The Significance of Cyberspace in Canadian Security Policy. *SOURCE Central European Journal of International & Security Studies* 7 (2): 102.

Leuprecht, Christian, Joseph Szeman, and David B. Skillicorn. 2019. The Damoclean Sword of Offensive Cyber: Policy Uncertainty and Collective Insecurity. *Contemporary Security Policy* 40 (3): 382–407.

Levin, A., and P. Goodrick. 2013. From Cybercrime to Cyberwar? The International Policy Shift and Its Implications for Canada. *Canadian Foreign Policy Journal* 19 (2): 127–143.

Mälksoo, M. 2018. Countering Hybrid Warfare as Ontological Security Management: The Emerging Practices of the EU and NATO. *European Security* 27 (3): 374–392.

Malone, E.F., and M.J. Malone. 2013. The "Wicked Problem" of Cybersecurity Policy: Analysis of United States and Canadian Policy Response. *Canadian Foreign Policy Journal* 19 (2): 158–177.

McDowell, Nathan, Ethan Walker, and Matthew Meyers. 2019. Prominent Cybersecurity Issues for the United States: A Qualitative Assessment (Chapter). In *Global Intelligence Priorities (from the Perspective of the United States)*. Hauppauge, NY: Nova Science Publishers.

McGuffin, Chris, and Paul Mitchell. 2014. On Domains: Cyber and the Practice of Warfare. *International Journal* 69 (3): 394–412.

NIS. 2019. National Intelligence Strategy of the United States of America. https://www.dni.gov/files/ODNI/documents/National_Intellige nce_Strategy_2019.pdf. Accessed 18 June 2019.

Norris, Donald F., Laura Mateczun, and Anupam Joshi. 2019. Cyberattacks at the Grass Roots: American Local Governments and the Need for High Levels of Cybersecurity. *Public Administration Review* 79 (6): 895–904.

NSS. 2017. National Security Strategy of the United States of America. http://nssarchive.us/wp-content/uploads/2017/12/2017.pdf. Accessed 23 Dec 2017.

Rudner, Martin. 2013. Cyber-Threats to Critical National Infrastructure: An Intelligence Challenge. *International Journal of Intelligence and Counterintelligence* 26 (3): 453–481.

Weaver, John M. 2017. Cyber Threats to the National Security of the United States: A Qualitative Assessment (Chapter). In *Focus on Terrorism*, vol. 15. Hauppauge, NY: Nova Science Publishers.

Weaver, John M., with Benjamin Johnson. 2020. *Cyber Security Challenges Confronting Canada and the United States*. New York, USA: Peter Lang.

CHAPTER 2

Research Questions, Methodology, and Limitations

Abstract This chapter provides an overview of the two research questions that will be used throughout the book. Moreover, it looks at the two models used to analyze data and explains the research limitations gravitating around secondary data.

Keywords YIRTM-C (M) · Federal secondary data case study triangulation model · TTPs

After reviewing the literature, the effects of several variables, most notably diplomacy, information, military, and economic (D.I.M.E.) means were used and studied; the potential impact of these on a shaped outcome of a weakened U.S. position was considered. Specifically, the author investigated the following research questions to look at what is possible from an adversarial perspective regarding cyber exploitation and attack.

Q1: How are state or non-state actors using cyber capabilities in the context of the York Intelligence Red Team Model-Cyber (Modified) [YIRTM-C (M)] in Fig. 2.1, to weaken the position of the United States?

J. M. Weaver, *The U.S. Cybersecurity and Intelligence Analysis*, https://doi.org/10.1007/978-3-030-95841-1_2

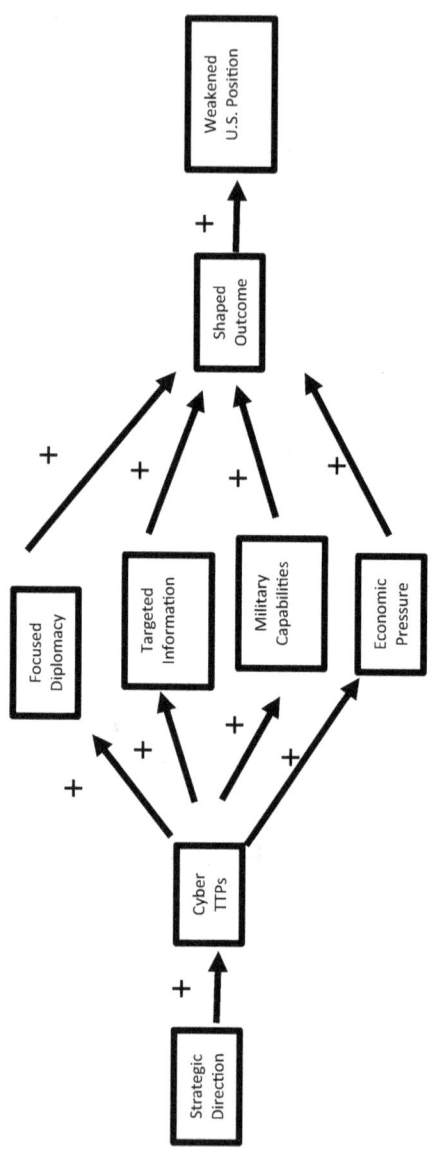

Fig. 2.1 YIRTM-Cyber (Modified)

Q2:   Why are state or non-state actors using cyber capabilities in the context of the YIRTM-C (M) to weaken the position of the United States?

Logic Model

The instruments of national power have been used by senior leaders historically to influence other countries and have been taught throughout the mid and senior-level staff and war colleges in the United States; this is done to foster an appreciation for making use of all available resources when attempting to shape outcomes and project power around the planet (JP 1–02 2010, 112). Conversely, this research considers the use of diplomacy, information, military, and economic means by states and non-state actors against the United States, essentially reversing the directionality of the concept to see how and why potential antagonists could use the instruments against the United States to weaken its security and position. Cyber tools, therefore, help enable states and non-state actors that do not otherwise enjoy a level of parity with the United States the opportunity to create a degree of asymmetry in terms of offensive capabilities. Those charged with the security of their country should afford consideration not just by viewing how they intend to apply pressure outward but must understand how those that want to harm this country will do so. To help make sense of what is occurring, this author used a modified logic model predicated on the four instruments of national power. Other publications have used D.I.M.E. [as the primary components of the YIRTM-C (M)] as a way to analyze threats (Weaver 2015; Weaver2 2015). Mattern et al. (2014) further underscore several of these issues in their research on the operational level of cyber intelligence.

The author used the York Intelligence Red Team Model-Cyber (Modified) [YIRTM-C (M)] to visually depict intervening relationships, thus assisting the researcher in the conduct of qualitative analysis. Similarly, as with directionality, the ordering of variables temporally is important (Weaver 2019). Thus, the ordering of items is useful, but not the sole consideration for helping to ground one's understanding of relationships using the instruments to shape outcomes. Regarding this research, the ultimate effect pursued by state or non-state actors through cyber exploitation or attacks is to weaken the position of the United States on the global stage, both materially and ideationally.

This model begins with the strategic direction of the actor that wishes to affect the national security of a country. The process starts with direction and guidance by a leader (either from state or non-state actors). With the direction and guidance of the leader of state and non-state actors, one can thereby focus on cyber and its use of the four instruments to put into motion tactics, techniques, and procedures (TTPs) supportive to a malicious actor's cause.

The first instrument of power is *diplomacy*. This instrument centers on leaders that engage others to foster favorable conditions to pursue their cause (Weaver and Pomeroy 2016, 2018, 2019; Weaver 2019). Diplomacy includes not just the application to one specific country but also looks regionally and/or globally at effects that decisions will have in contiguous countries and beyond to undermine the advancement of the interests of the United States. Diplomacy is often understood to be far less costly in terms of short-term expenditures by developing those relationships to foster the pursuit of one's position.

Information is the second instrument and is often equated to power and its applied use can effectively influence events throughout the world (Weaver and Pomeroy 2016, 2018, 2019; Weaver 2019). Targeted messaging can be very useful in shifting public opinion to help promote one's cause while simultaneously delegitimizing the public relations campaign of another country or non-state actor (Weaver and Pomeroy 2016, 2018, 2019; Weaver 2019). Adversaries can make use of information primarily through social media conduits to inexpensively 'educate' others to their objectives to promote greater support as the non-state actor or country enhances its position. Information acquisition is also a precursor of success. Here an actor would follow through with efforts to acquire and exploit data from social media, bulk information stored in databases by large telecommunications centers, records maintained by the finance sector, and other sources to implement cyber espionage to cause instability or worse. This instrument, interestingly, is often considered the least expensive when compared to the other instruments of power and it requires little in terms of monetary resources (albeit training is a necessity as a precursor). Information as an instrument of power is seemingly available and often easy to acquire by all countries and non-state actors alike.

Of the instruments of power, conventionally, the one most understood by defense establishments and the senior leadership of a country's government is the *military* (Weaver and Pomeroy 2016, 2018, 2019; Weaver 2019). Under regular circumstances, the use of this element of power might include humanitarian relief events, 'show of force' exercises, the deployment of peacekeeping forces, peace enforcement missions, and full-up combat operations. In the context of an adversarial power trying to leverage cyber, one might look to the horizontal and vertical linkages of military capabilities and command, control, communications, computers, intelligence, surveillance, and reconnaissance (C4ISR) systems to cyber-attacks. During the month of July in 2019, the U.S. Defense Department released its digital modernization strategy focusing on pursuing greater resiliency of the military network (DOD DMS 2019). Though the U.S. still arguably maintains the most formidable and robust military capability in the eyes of the world, much of its combat equipment is reliant on the interoperability of those C4ISR systems, and therefore, they tend to be inherently vulnerable to CNE and CNA.

Like military power, the last instrument—*economic*—represents a conventional means of control and influence. Possession of money has often been seen as a precursor to power (Weaver and Pomeroy 2016, 2018, 2019; Weaver 2019). Accordingly, nations have used their economic power to influence outcomes. Under conventional conditions, economically, countries can pressure others by using the 'power of the purse' coercively to foster behavioral changes that are in line with one's objectives.

However, purveyors of cyber TTPs are often looking for something different. More specifically, through data collection (and more pointedly, the informational component), one can make use of malware, altered data, or viruses to destabilize financial sectors, wreak havoc on electrical power grids, or possibly to manipulate official statements or messages (and use 'deep fakes') to either weaken the position of the United States, fostering greater distrust, or a lack of confidence from within the country (of the United States' power) and even its reputation on the world stage.

Methodology

As a result of the literature review, it is apparent that only one other research project looked at 'red teaming' the application of the instruments of national power against the United States (Weaver and Johnson 2020). This research looks to expand on that approach to fill the gap focusing specifically on cyber issues and looking at what *is* happening and what *could* happen by conducting a deeper dive using a series of micro-case studies looking at four nations (China, Iran, North Korea, and Russia) and two non-state actors (al Qaeda and the Islamic State). Research data used in this book emanates from secondary open-source (unclassified) data. Moreover, a specific approach to triangulation was employed to understand how and why specific cyber incidents have occurred. Specifically, this book looks at cyber threats and what has been done by the selected state and non-state actors using a new (albeit modified) analytical model, the YIRTM-C (M), to triangulate this data and offer an analysis of specific actors along with their cyber capabilities and intentions concerning undermining the security of the United States.

Contemporary research has underscored secondary data viability (Weaver and Pomeroy 2016, 2018, 2019; Weaver 2019; Weaver and Johnson 2020). Since this type of data is relatively inexpensive and readily available, practitioners and researchers alike in the social science and policy fields have turned to these sources specifically to leverage the information to conduct research. This research only makes use of secondary data to see what is occurring; this type of data has an advantage in cross-sectional designs. Future research can tailor mainstream primary data collection techniques to build on the findings of this study.

The author selected a model previously used in other peer-reviewed research to help ensure a balanced approach to problem dissection (Weaver and Johnson 2020). The methodological approach selected was the Federal Qualitative Secondary Data Case Study Triangulation Model (FQSDCS) depicted in Fig. 2.2 (Weaver3 2015; Weaver and Pomeroy 2016, 2018, 2019; Weaver 2019; Weaver and Johnson 2020).

This approach to triangulation consists of three concentric components to arrive at results (Weaver and Pomeroy 2016, 2018, 2019; Weaver 2019). The first of these includes *plans* and *systems* (Weaver and Pomeroy 2016, 2018, 2019; Weaver 2019). *Plans* and *systems* turn to what is in existence at present and the premise focuses on what TTPs have been used by others thus far against the United States. This component makes

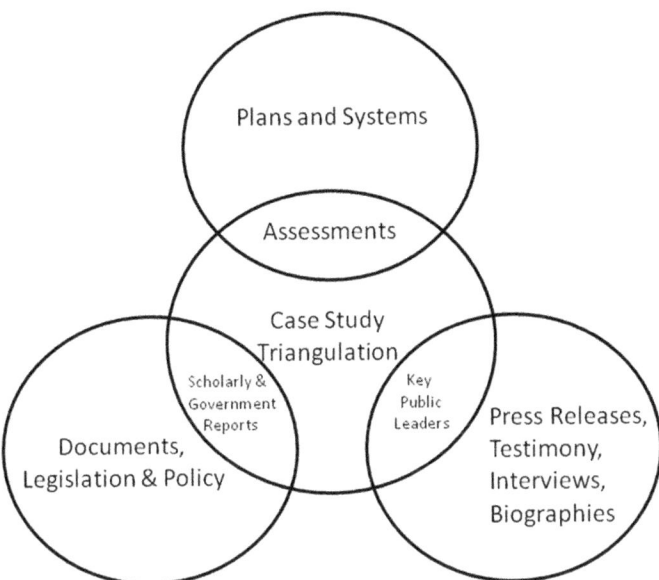

Fig. 2.2 Federal qualitative secondary data case study triangulation model

use of secondary data sources to assess what has been identified in the United States that is vulnerable to cyber as important to protect against CNO, CNE, and CNA as well as information on threats to the national security of the United States. Systems afford consideration to capabilities and/or procedures in existence to either initiate CNO to protect against CNE and CNA efforts. A subset of this component looks to assess the outcomes achieved at present and looks at the continuance of vulnerabilities that exist with regard to the infrastructure in the United States. Through this first component of the model, one can infer what offensive cyber capabilities may be employed by state and non-state actors against the United States as well as what security measures exist or are being developed by the United States to protect against existential cyber threats.

The model's second component gravitates on *documents, legislation, and policy* to better understand the threat and the various types of exploitation and attacks that have occurred in this country (Weaver and Pomeroy 2016, 2018, 2019; Weaver 2019). This component affords consideration to other written works and focuses on peer-reviewed journals and government reports primarily published during the last seven years to better understand the current security environment with a focus on cyber threats. Generally, this component of the Federal Qualitative Secondary Data Case Study Triangulation Model forgoes mainstream news reports and turns to more credible sources of information which is subsequently used as raw data. More specifically, this study made use of a multitude of sources which include reports generated by the Government Accountability Office, the Congressional Research Service, books on cyber, the Congressional Budget Office, as well as peer-refereed articles such as those published in the *International Journal of Intelligence and Counterintelligence*. These sources, therefore, provide an understanding of current capabilities and threats confronting the United States and the subsequent assessment at the end of this book looks at synthesizing these themes.

The final component of the model focuses attention primarily on oral accounts to balance the triangulation process (Weaver and Pomeroy 2016, 2018, 2019; Weaver 2019; Weaver and Johnson 2020). This component includes official press releases denoting exploitation attempts and attacks, public interviews, testimony (before government bodies), and biographical accounts. Moreover, the process affords consideration to key leader statements and the messaging coming from those both inside the government of the United States, and other prominent figures from potential nefarious states and non-state actors. Specific sources include official transcripts of testimony by cyber experts, interviews of prominent cyber experts, intelligence careerists, and defense sector employees, in addition to the official press releases by defense personnel and professionals working in the Intelligence Community.

All scholarly research must achieve validity. Accordingly, this research addressed validity by considering two areas. One, variable checks were implemented. These included considerations of efforts to verify that the variables measured what they were supposed to measure. Likewise, construct validity procedures in this work made use of peer-reviewed articles to corroborate findings in this study.

Next, face validity strives to ensure whether something intuitively makes sense (Creswell 2008). Moreover, face validity techniques in this book evaluated the logic model, variable directionality, and its purpose to more aptly ensure that what was presented was not counterintuitive to what one would reasonably consider when looking at impacts. As was stated previously in the methodology section, this author used the logic model to frame the issue more directly as to how both state and non-state actors might consider using the instruments of national power against the United States to weaken its strategic position on the global stage and to determine why. More pointedly, the Federal Qualitative Secondary Data Case Study Triangulation Model, vetted and used in other research, allowed this researcher to ensure a balance of secondary sources to arrive at more robust findings than would otherwise be possible.

Emphasis was also ascribed to the issue of reliability. The researcher utilized an approach that was consistent with research endeavors to contribute to reliability. Likewise, it is necessary for consistency in measurement when looking to achieve reliability in a study when conducting research. To facilitate this, the study was underpinned by the consistency of the D.I.M.E. instruments of power through the use of the YIRTM-C (M) and the Federal Qualitative Secondary Data Case Study Triangulation Model to analyze the raw data (Weaver 2019). Multiple sources were therefore used to corroborate information.

This research used the YIRTM-C (M) and FQSDCS models as an approach to analyzing cyber threats and cyber issues. The understanding generated through the application of the models will better facilitate answers to the two main research questions presented earlier in this chapter. Moreover, it is incumbent upon a researcher to achieve a greater understanding of the various cyber activities taking place on the planet, and the researcher looked to see who is most likely behind these nefarious activities, why they are occurring, and what might unfold in the immediate future for cybersecurity to mitigate cyber threats (Mattern et al. 2014, 704).

LIMITATIONS

Studies using this type of information are limited in such areas as the variables explored and the time horizon considered, as is the case with research primarily centered on secondary data (Weaver and Pomeroy 2016, 2018, 2019; Weaver 2019; Weaver and Johnson 2020). Broadly, the scope of this research was limited to seven years though there were some instances dating back a few more years. This researcher ascribed greater weight to data in the latter years because it was more relevant as it most likely pertains to the future and because of the ever-changing situation on the planet in a globalized setting. Though the variables in the

YIRTM-C (M) were selected, there are other factors that exist that could impact the true value of each (Weaver 2019).

This study only focused on intervening relationships. It is probable that other moderators could have an influence on the findings' interpretation (Weaver and Johnson 2020). These include such areas as legacy trade agreements, treaties, the level of sophistication of technology, and more. Follow-on studies might afford consideration to these to build upon the results of this study.

Likewise, in the case of all qualitative research, one cannot generalize beyond the specific study. The analysis of data and the results thereof focused only on the use of cyber as it applies to the specific state and non-state actors mentioned previously.

Other weaknesses from secondary data research emerge from the cross-sectional design. Due to the reliance solely on secondary sources, it is difficult for one to know if bias existed from how the data was originally collected (Cross-Sectional 2012). Challenges also can arise regarding the process of using a cross-sectional design when interpreting results based in part on bias issues and its lack of full inclusion of all events (Cross-Sectional 2012).

The author did not consider primary sources. The primary research techniques include such things as personal observations, interviews, and surveys. Though excluded from this research endeavor, future studies might consider these techniques to either confirm results or to complement the findings.

One must understand that no research is perfect and complete, but results can still be useful. Moreover, as additional information on cyber TTPs emerge and become available and as global circumstances change, the results found in this book can be used in confirmatory research or as a contributor to efforts to help expand on threat understanding linked to cyber conveyances in the context of the YIRTM-C (M). The author performed a meta-analysis to find data. More to the point, the keywords that were used follow in Annex 2.1 (and this also includes synonyms).

Annex 2.1: Meta-Analysis (words)

Advanced Persistent Threats (APT)
Al Qaeda
Asymmetric Warfare
Blockchain

Bots
China
Computer Network Attack (CNA)
Computer Network Defense (CND)
Computer Network Exploitation (CNE)
Computer Network Operations
Cryptocurrency
Cyber
Cyber Activities
Cyber-attack
Cyber Capabilities
Cyber Reconnaissance
Cyber Threat Intelligence
Cyberspace
Cybersecurity
Cyber Threats
Cyber Threat Intelligence
Cyber-Warfare
Dark Web
Deep Fakes
Denial of Service
Diplomacy
Economic
Espionage
G7
G8
G20
Government
Government Report
Hybrid Warfare
Humint
Information
Intelligence
Internet
Interview
IoT
Iran
Islamic State
IT

JCPOA
Journal
Kubernetes
Legislation
Malware
Nafta
NATO
Network
North America
North Korea
Military
Plan
Power
Press Release
Ransomware
Reconnaissance
Russia
SCADA
Sigint
Spear-phishing
System
TOR
Trade
United States
Usmca
VPN
Yirtm

References

Creswell, J.W. 2008. *Research Design: Qualitative, Quantitative, and Mixed Methods Approaches*. Thousand Oaks, CA: Sage.
Cross-Sectional. 2012. *Cross-Sectional Studies*. http://www.healthknowledge. org.uk/public-health-textbook/research-methods/1a-epidemiology/cs-as-is/ cross-sectional-studies. Accessed 17 Oct 2012.
DOD DMS. 2019. DoD Digital Modernization Strategy. https://media.def ense.gov/2019/Jul/12/2002156622/-1/-1/1/DOD-DIGITAL-MOD ERNIZATION-STRATEGY-2019.PDF. Accessed 21 July 2019.

JP 1–02. 2010. *Department of Defense Dictionary of Military and Associated Terms. Joint Publication 1–02.* November 8, 2010.

Mattern, Troy, John Felker, Randy Borum, and George Bamford. 2014. Operational Levels of Cyber Intelligence. *International Journal of Intelligence and Counterintelligence* 27 (4): 702–719.

Weaver, John M. 2015. The Perils of a Piecemeal Approach to Fighting ISIS in Iraq. *Public Administration Review* 75 (2): 192–193.

Weaver2, John M. 2015. The Enemy of My Enemy Is My Friend…Or Still My Enemy: The Challenge for Senior Civilian and Military Leaders. *International Journal of Leadership in Public Service* 11 (3–4).

Weaver3, John M. 2015. The Department of Defense and Homeland Security Relationship: Hurricane Katrina Through Hurricane Irene. *Journal of Emergency Management* 12 (3): 265–274.

Weaver, John M. 2019. *United Nations Security Council Permanent Member Perspectives Implications for U.S. and Global Intelligence Professionals.* Peter Lang Publishing.

Weaver, John M., and Benjamin Johnson. 2020. *Cyber Challenges Confronting Canada and the United States.* New York, USA: Peter Lang Publishing.

Weaver, John M. (Editor) with Jennifer Pomeroy (Editor). 2016. *Intelligence Analysis: Unclassified Area and Point Estimates (and Other Intelligence Related Topics).* Nova Science Publishers.

Weaver, John M. (Editor) with Jennifer Pomeroy (Editor). 2018. *Intelligence Analysis: Unclassified Area and Point Estimates (and Other Intelligence Related Topics) 2nd Edition.* Nova Science Publishers.

Weaver, John M. (Editor) with Jennifer Y. Pomeroy (Editor). 2019. *Global Intelligence Priorities (from the Perspective of the United States).* Nova Science Publishers.

CHAPTER 3

The United States

Abstract This chapter provides a brief history of the United States. Likewise, it goes into detail on the national security priorities of this country.

Keywords National security · Homeland · Intelligence Community · Hard power · Soft power

OVERVIEW OF THE UNITED STATES

The United States is a constitutional republic and the American colonies severed ties with Britain in 1776 and officially became the United States of America following the implementation of the *Treaty of Paris* in 1783 (CIA 2019). This country saw an increase of 37 new states that were added to the original 13 as the nation expanded westerly across the North American continent and it acquired several overseas possessions throughout the nineteenth and twentieth centuries. Two of the most significant events in this nation's history were the Civil War (it took place from 1861 to 1865), in which the northern states (Union) fought with and subsequently ended up defeating a secessionist Confederacy of 11 southern slave-owning states (CIA 2019). The second was the Great

© The Author(s), under exclusive license to Springer Nature Switzerland AG 2022
J. M. Weaver, *The U.S. Cybersecurity and Intelligence Analysis*,
https://doi.org/10.1007/978-3-030-95841-1_3

Depression that occurred in the 1930s, which saw a significant economic downturn during which about a quarter of the labor force lost its jobs (CIA 2019). With the victories of World Wars I and II and the ending of the Cold War in 1991, the United States remains the most powerful nation in the world. Since the end of World War II, the economy of this country has achieved relatively low unemployment, steady growth, generally controlled inflation, and is rapidly burgeoning in technological innovation (CIA 2019).

The United States' capital is in Washington, the District of Columbia. The country is comprised of 50 states and one district; these include Alabama, Alaska, Arizona, Arkansas, California, Colorado, Connecticut, Delaware, Florida, Georgia, Hawaii, Idaho, Illinois, Indiana, Iowa, Kansas, Kentucky, Louisiana, Maine, Maryland, Massachusetts, Michigan, Minnesota, Mississippi, Missouri, Montana, Nebraska, Nevada, New Hampshire, New Jersey, New Mexico, New York, North Carolina, North Dakota, Ohio, Oklahoma, Oregon, Pennsylvania, Rhode Island, South Carolina, South Dakota, Tennessee, Texas, Utah, Vermont, Virginia, Washington, West Virginia, Wisconsin, and Wyoming (CIA 2019). The United States also has many territories, and these include the following: American Samoa, Baker Island, Guam, Howland Island, Jarvis Island, Johnston Atoll, Kingman Reef, Midway Islands, Navassa Island, Northern Mariana Islands, Palmyra Atoll, Puerto Rico, Virgin Islands, and Wake Island. The United States achieved independence from Great Britain on July 4, 1776.

Under the executive branch, this nation's chief of state is the president and Joe Biden has been president since January 20, 2021. It is important to realize that the president is not only the chief of state but also the head of government as well (CIA 2019). The cabinet is appointed by the president but those occupying the positions must be approved by the Senate. In the United States, under elections and appointments, the president and vice president are both indirectly elected on the same ballot by the Electoral College of 'electors' chosen from each state; accordingly, the president and vice president usually serve a 4-year term and are generally eligible for a second term (CIA 2019).

The second branch of government, referred to as the legislative branch, is a bicameral Congress that consists of a Senate (with 100 seats, two members are directly elected in each of the 50 state constituencies by a simple majority vote with two exceptions: Georgia and Louisiana that require an absolute majority vote with a second-round if necessary) (CIA 2019). Members thus serve 6-year terms whereby one-third of member-ship positions are up for election every two years. The legislative branch's second component is the House of Representatives; it comprises 435 seats where its members are directly elected in single-seat constituencies by a majority vote (there is one exception: Georgia which requires an abso-lute majority vote with a second-round if necessary) (CIA 2019). These members end up serving for two years.

The judicial branch is the third component of the United States' government. The highest court, the U.S. Supreme Court, is composed of nine justices—the chief justice and eight associate justices (CIA 2019). The president has the power to nominate and, with the advice and consent of the U.S. Senate, subsequently appoints Supreme Court justices. Justices end up serving for life. The subordinate courts at the federal level include the Courts of Appeal (made up of the U.S. Court of Appeal for the Federal District and 12 regional appeals courts) and 94 federal district courts in all 50 states and territories (CIA 2019).

Economically, the United States ranks first on the global stage. Accord-ingly, the U.S. possesses the most technologically powerful economy in the world, with a per capita GDP of \$59,500 (CIA 2019). Companies in the United States are often at or near the forefront regarding tech-nological advances; this is especially true in the industries pertaining to computers, aerospace, medical, pharmaceuticals, and military equipment (CIA 2019). That stated, the advantage has narrowed markedly from the end of World War II. Based on GDP comparison, and as measured by purchasing power parity conversion rates, the United States' economy in 2014, having worked its way to the largest in the world for more than a century, fell into second place behind China; the PRC has more than tripled the U.S. growth rate for each year of the past four decades (CIA 2019).

Private business firms and individuals in the United States make most of the decisions, and the governments at the federal and state levels buy needed goods and services predominantly from the private marketplace. Business firms in this country often enjoy greater flexibility than countries in Japan and Western Europe when looking at decisions to develop new products, expand capital plants, and whether to lay off surplus workers. Simultaneously and historically, businesses in the U.S. end up facing higher barriers to enter their rivals' home markets than foreign firms face entering U.S. markets.

Longitudinally, problematic issues for this nation include such things as stagnation and falling wages for lower-income workers, inadequate investment in deteriorating infrastructure, and the rapid rise in medical and pension costs of an increasingly aging population (CIA 2019). The United States has also seen a significant budget deficit, sporadic energy shortages, and rising national debt (CIA 2019).

Though the proliferation of technology has been a key underpinning in the incremental development of a "two-tier" labor market whereby those at the bottom oftentimes lack skills and education, and the technical and professional skills of those at the top and, more and more, are often unable to earn comparable health insurance coverage, pay raises, and other defined benefits (CIA 2019). However, through the globalization of trade, and especially when viewed in the context of rising low-wage producers such as China, these have put additional emphasis on wage suppression and upward pressure on the return of capital (CIA 2019). Since the middle of the 1970s, practically all the gains in household income have been seen by the top 20% of households. Moreover, since 1996, capital gains and dividends have grown faster than wages or any other category of after-tax income (CIA 2019).

Consider the impact of oil; this commodity accounts for more than 50% of U.S. consumption, and oil as a resource has a major impact on the overall health of this economy (CIA 2019). Between 2001 and 2006 alone, crude oil prices have doubled, the latter being the year home prices peaked. What's more, is that rising gasoline prices chipped away into consumers' budgets and, as a result, many individuals fell behind in their mortgage payments. Oil prices skyrocketed between 2006 and 2008 by another 50%, and what ensued saw home foreclosures more than double during that two-year period (CIA 2019). Besides the weakening of the housing market, rising oil prices caused a significant drop in the value of the U.S. dollar and a deterioration in the U.S. commercial trade deficit that ended up peaking at $840 billion in 2008. As a result of the U.S. economy being intensively energy-dependent, falling oil prices since 2013 and continuing until recently have alleviated most of the challenges and issues earlier increases had brought to bear (CIA 2019). That stated, oil prices have soared in 2021.

More pointedly, the subprime mortgage crisis, freefalling home prices, investment bank failures, tight credit, and the global economic downturn pushed the United States into a recession by mid-2008 (Weaver and Johnson 2020). This country's GDP contracted until the third quarter of 2009 and resulted in the deepest and longest downturn since the Great Depression of the 1930s (CIA 2019). To help stabilize the financial markets, the U.S. Congress created a $700 billion Troubled Asset Relief Program (TARP) in October 2008 (CIA 2019). Accordingly, the government used some of this currency to purchase equity in U.S. industrial corporations and banks, much of which had been repaid to the government by early 2011 (Weaver and Johnson 2020).

Congress passed, and former President Barack Obama signed a bill into law in early 2009 that allocated an additional $787 billion fiscal stimulus that was to be implemented over 10 years; 2/3rds was for additional spending and 1/3rd was spent on tax cuts (CIA 2019). The stated purpose was to (1) create jobs and (2) help with economic recovery (Weaver and Johnson 2020). What ensued in 2010 and 2011 was the federal budget deficit of this country reaching nearly 9% of GDP (CIA 2019). In the year that followed, the federal government reduced the growth of spending and the deficit contracted to 7.6% of GDP. The manifestation of this was that U.S. revenues from taxes and other sources were then lower (as a percentage of GDP) than those of most other nations (CIA 2019).

The War on Terror and more specifically, the operations in Iraq and Afghanistan required major shifts in the allocation of national resources from civilian to military purposes and this significantly contributed to the rise of both the budget deficit and public debt (CIA 2019). Through fiscal year 2018, the direct costs tied to the wars have totaled more than $1.9 trillion, according to U.S. government figures (Weaver and Johnson 2020). Moreover, COVID stimulus and the new infrastructure bill in this country have exponentially added to the debt of the United States as of 2021.

Other related costs and specifically looking at the field of health care, in March 2010, former President Obama signed into law the Affordable Care Act (ACA) (CIA 2019). This initiative was a major health insurance reform that was expressly designed to expand coverage to an additional 32 million Americans by 2016, through privately owned health insurance for the general public and Medicaid for those that the government determined to be impoverished. Total spending on the Affordable Care Act, which includes both public plus private costs, rose from 9.0% of GDP in 1980 to 17.9% in 2010 (CIA 2019).

In December 2017, Congress passed and subsequently, President Donald Trump signed the Tax Cuts and Jobs Act into law, which, among multiple provisions, markedly reduced the corporate tax rate from 35 to 21%; likewise, it lowered the tax rate for those individuals with the highest incomes by more than 2% from 39.6 to 37%, and by lower percentages for those at the bottom tier of income levels (CIA 2019). This changed many deductions and credits used to calculate taxable income and subsequently eliminated the penalty imposed on taxpayers in 2019 for those who did not obtain the minimum amount of health insurance required under the ACA. The new tax structure took effect beginning in January 2018 (CIA 2019). The tax cuts for corporations are supposed to be permanent; conversely, those for individuals are scheduled to expire after 2025. Under the Congressional Budget Office, the Joint Committee on Taxation (JCT) later estimated that the new law would see a reduction in tax revenues and therefore increase the federal deficit by approximately $1.45 trillion over the 10-year period between 2018 and 2027. This amount would decline, however, if the economy's growth were to exceed the JCT's prediction (CIA 2019).

The United States possesses a formidable military. United States' Department of Defense consists of the U.S. Army, U.S. Navy (including the Marine Corps), U.S. Air Force, U.S. Space Force, and during times of war, the U.S. Coast Guard (during peacetime, the Coast Guard falls under the Department of Homeland Security, however in wartime falls under the Department of the Navy). The United States' spending on the North Atlantic Treaty Organization (NATO) represents 69.3% of the total share contributed by all NATO member nations (NATO 2018, 35).

The United States' Intelligence Community has 17 subordinate organizations under the Office of the Director of National Intelligence (DNI 2019). More pointedly, it comprises two independent agencies: The Office of the Director of National Intelligence (ODNI), and the Central Intelligence Agency (CIA) (DNI 2019). Defense Department controls nine of the agencies; apart from each of the branch's intelligence components (Army, Navy, Air Force, Marine Corps, and Space Force), these also include the Defense Intelligence Agency (DIA), the National Security Agency (NSA), the National Geospatial-Intelligence Agency (NGA), and the National Reconnaissance Office (NRO). The last three of which are very cyber-centric (Lowenthal 2017; DNI 2019). There are seven other members; the most important include the State Department's Bureau of Intelligence and Research, the Department of Homeland Security's Office of Intelligence Analysis, and the intelligence component of the Federal Bureau of Investigation (DNI 2019).

President Biden issued his Interim National Security Strategic Guidance (INSSG) on March 3, 2021 (INSSG 2021); as of publication, the NSS of 2017 is still what is technically in effect. This capstone document serves as the azimuth for prominent government leaders to guide them through the implementation of core duties at federal agencies and departments (Weaver 2018). Though the INSSG is only 23 pages long, it outlines the host of issues that he and by extension, his National Security Council, sees as concern for the United States. That stated, the most recent official NSS looks to (1) protect the homeland, (2) foster the promotion of the prosperity of the U.S., (3) leveraging national strength to preserve peace, and (4) advance U.S. influence throughout the world (NSS 2017, 4). Likewise, the National Defense Strategy and U.S. Department of State's Joint Strategic Plan came out early in 2018 less than two months after the release of the most recent NSS.

The Homeland

The NSS identifies key threats to the United States, and these include a multitude of issues, like North Korea and its pursuit of weapons of mass destruction (WMD) capable of reaching the United States (Wright 2009, 5; INSSG 2021). Other issues include Iran's support of terrorist groups, jihadist terror organizations whose intent is bent on targeting the American people, and threats from cyber (NSS 2017, 7; INSSG 2021).

Moreover, the NSS affords consideration to a series of priority risks and these include national security, safety and health, finance and banking, communications, energy, and transportation (NSS 2017, 13). Accordingly, the U.S. should invest in the protection of its infrastructure and hardening of key targets to make these less susceptible to physical attacks, especially ports (air and sea), railways and roads, the telecommunications industry, in addition to transit systems (NSS 2017, 19; INSSG 2021). Other focal points identified in the document include sealing off porous entry points into the United States. The United States' Joint Strategic Plan (authored by the Department of State) also underscores the importance of security at home through its Joint Strategic Plan or JSP (JSP 2018, 23). The Defense Department plainly states that the homeland will not be a sanctuary for terror organizations. The United States, however, remains a target with exposed flanks in such areas of cyber threats, possible future attacks on its citizens, and more (NDS 2018, 3). The federal government is a critical level of government that must work with state and local governments to identify cyber threats and to help ensure adequate resiliency measures are built into networks to mitigate exposure from state and non-state nefarious types alike especially for transportation, communication networks, the financial and banking sector, and the health industry. The United States views threats through the lens of cyber hackers, malevolent non-state actors, and more (NDS 2018, 3; INSSG 2021).

The U.S. will promote strong diplomatic initiatives with China, South Korea, and Japan to show regional resolve regarding North Korea to help protect its homeland and will also be a key actor garnering support among the other permanent members of the United Nations Security Council in seeking to weaken Kim Jong-un's regime (Weaver 2018, 65). The United States will be involved in implementing a layered missile defense system and will be called to pursue transnational terror threats at their source (NSS 2017, 8–10). It is of paramount concern that the U.S. government work with state and local governments to identify cyber threats and to help ensure the resiliency of networks to exposure from the likes of state and non-state nefarious types and to share information with key allies (and by extension, their intelligence services) throughout the globe (NSS 2017, 13).

Prosperity

The U.S. will have major challenges as it promotes American prosperity. Because of this, it will be necessary to pursue trade agreement negotiations that are seen as more favorable to the United States; this comes at a period when U.S. growth has been averaging a paltry 2% annually over recent years compared to China's double-digit numbers. Likewise, the United States will want to ensure sea lanes remain open to foster free trade especially in the South China Sea where China has become more aggressive with its influence in recent years. The National Defense Strategy underscores just how China is engaged in predatory economics and other influence operations with the express intent of bolstering its prosperity (NDS 2018, 2).

Peace Through Strength Abroad

The U.S. has attested to the need for preserving peace through strength showing the U.S. as still being a relevant global power (NSS 2017, 23; INSSG 2021, 6). In doing so, President Biden identified China and Russia as major challengers to the United States (INSSG 2021, 6). The U.S. will also turn to its diplomatic cadre to strengthen relations and alliances in Europe and Asia (for military basing rights, trade, and more) while looking to engage both countries to find common ground to help turn around the deterioration in the situation seen recently regarding both. The U.S. will likely continue military exchanges and participate in various show of force exercises to demonstrate U.S. capabilities and resolve.

Biden has concerns about Iran's support for and sponsorship of terror organizations and the possibility that it might renege on its agreement to halt uranium enrichment (INSSG 2021). Accordingly, the United States must lean on the United Nations and the International Atomic Energy Agency (IAEA) to keep pressure on Iran to compel it to move away from its support of terror organizations and from it returning to pursuing a nuclear weapons' development program especially as the U.S. has pulled out of the nuclear agreement and as it possibly looks to renegotiate new terms. The Defense Department of the United States views Iran as a regional actor and one that has intentions of becoming a major power in the Middle East and sees it as a state sponsor of terrorism in the region (NDS 2018, 2).

The NSS (2017) further states that North Korea has heavily invested in a ballistic missile program and has amassed a significant nuclear capability (and arsenal). The Defense Department even went as far as to consider North Korea as a rogue regime actively trying to destabilize the region of Northeast Asia (NDS 2018, 2). Even the U.S. State Department is squarely behind the idea of countering the proliferation of weapons of mass destruction to include those from this regime (JSP 2018, 23). Aside from what was covered about this Asian antagonist in the Homeland section of the NSS, the U.S. will have a role regionally in maintaining a formidable forward military presence in South Korea, Japan, and Guam and through the provision of missile defense capabilities in Northeast Asia for the foreseeable future. This shows both commitment and resolve to allies of the United States.

Jihadists operating throughout the globe present another major concern. As Biden, and by extension his federal departments and agencies, looks to continue to implement the INSSG, the administration needs to figure out how the U.S. will apply economic pressure to cut off funding, look to use the military kinetically to take out terror command and control centers of gravity, and turn to the Department of State to follow through with other capabilities (diplomatic, military and economic) of friends and allies to collectively pursue targets while simultaneously looking to help defeat the ideology through a successful campaign using information to more aptly dispel the falsehoods of their messaging. What's more is that the State Department's JSP looks to leverage the strengths of its partnerships the world over to defeat the Islamic State, al Qaeda, and other terror organizations throughout the planet (JSP 2018, 25). Likewise, the United States can selectively distribute aid and support efforts to hopefully prevent countries from falling into the weak or failed state categories that could harbor transnational jihadists particularly by supporting governments that are allied with the United States.

Advancing American Influence

The United States wants to retain its position as a hegemonic power. Accordingly, it would like to counter emerging powers like China and Russia as they attempt to wrest influence and power from this nation. Likewise, the U.S. wants to reduce the dependency of other nations on the United States' foreign aid (NSS 2017, 39). The NSS also underscores the importance of leveraging the Department of State in creating trade agreements with countries in Africa, Asia, and Latin America (NSS 2017, 39). The U.S. plans to leverage technology to serve in a complementary role to diplomatic efforts that it is pursuing overseas and would like to support incentivizing reforms across the globe to include such organizations as the International Monetary Fund, World Bank, and World Trade Organization (NSS 2017, 40). Though the U.S. needs to remain the global leader (through the United States' vision), it wants to hold other nations accountable and to have them share the financial burden when solving complex problems.

This country has conducted a reinvestment campaign in its military in recent years as a way to reverse the atrophy of power that it has experienced in recent years (NDS 2018, 1). It has done this to stem the erosion of influence it has seen in current times to be able to project its influence worldwide to deter war and protect vital security interests of the country (NDS 2018).

Soft Power

The United States' State Department is resolved to help bring the NSS to fruition. To help realize it, this department executed its plan through four primary goals.

The first goal is to protect U.S. security at both home and abroad (JSP 2018, 23). To achieve this, it will pursue a series of strategic objectives. The first sub-goal involves efforts oriented to counter WMDs (JSP 2018, 23). The department will actively implement diplomatic engagement to leverage the strengths of partner nations (JSP 2018, 25). Likewise, it will work with other departments within the federal government of the U.S. to achieve synergy through organizations like the Department of Energy, Department of Justice, Department of Treasury, and Department of Commerce.

It hopes to defeat terrorist organizations like the Islamic State and al Qaeda under a separate sub-goal (JSP 2018, 24). Like goal 1.1, it intends to do this by working with U.S. federal agencies and departments while simultaneously seeking out assistance diplomatically from other countries' defense, law enforcement, and judicial sectors (JSP 2018, 26). It also will serve in a complementary role to the other major department that is involved in foreign engagement, the Department of Defense (JSP 2018, 27).

Soft power initiatives found under the first goal also look towards countering transnational crime and violence directed at U.S. interests (JSP 2018, 27). Likewise, it hopes to bring about success through strengthened citizen-responsive governance, fostering human rights, supporting democracy globally, and promoting security initiatives (JSP 2018, 29).

Under this goal of protecting America's security, it orients towards two additional sub-goals. These both include strengthening the resiliency of partners and allies and improving border security at home.

The JSP covers a second goal to help renew the competitive advantage for America linked to economic growth (JSP 2018, 23). Under this category, a sub-goal of the department is looking to leverage other international organizations and by implementing bilateral agreements that could result in commercial arrangements to help improve its economic position (JSP 2018, 23). What's more, is that the U.S. State Department is interested in opening global markets and pursuing initiatives that will lead to economic security and reforms in the areas of governance and economic initiatives.

There are other soft power initiatives. The JSP shifts then to its third major goal of promoting leadership in the world through balanced engagement (JSP 2018). To realize this, the department would like to pull back from aid distribution while it looks to foster the continuance of partnerships the world over. It wants to remain engaged through the pursuit of foreign policy initiatives that strive to balance the burden of world leadership with other nations simpatico with the values espoused by the United States (JSP 2018).

Hard Power

The U.S. Defense Department is poised to leverage the military instrument of power if and when needed. Moreover, it is committed to defending this country's homeland from attack (NDS 2018, 4). It is doing so through an investment in a ballistic missile defense system, employment of National Guard personnel along the southern border of the United States, and the construction of a wall.

This department makes use of the technological superiority of weapon systems and employs forces as necessary globally and specifically in key regions (NDS 2018, 4). The intent is to use stand-off security measures to minimize troop exposure where possible to threats from state and non-state actors alike. In doing so, it wants to expressly deter adversarial aggression that could be directed against national security interests.

More specifically, the arraying of the U.S. armed forces will occur in the Indo-Pacific region, on the European continent, the continuance of operations in the Middle East, and is rounded out by the security of the Western Hemisphere (NDS 2018). What is inextricably linked to this is the desire to defend allies of the United States from military aggression from nefarious actors (NDS 2018).

Prevention of hostilities is preferred over having to respond to acts of violence ex post facto. Accordingly, DOD remains committed to preventing terrorists from realizing success both at home and overseas (NDS 2018). Along the line of terrorism, efforts will include serving as a force to dissuade, prevent, and deter state and non-state actors from pursuing the acquisition and use of WMDs (NDS 2018).

To better realize the chance for success, the Defense Department will implement a strategic approach. This includes the use of the four instruments of power (identified earlier) and the understanding of the following linkages: finance, intelligence, and law enforcement (NDS 2018, 4). Likewise, it will continue to invest in a military modernization initiative. This includes the likes of the nuclear triad (sub-launched ballistic missiles, nuclear missiles, and nuclear bombs capable of being dropped from aircraft).

Investments will also encapsulate both space and cyberspace capabilities to operate in a myriad of domains (NDS 2018). Other initiatives will result in enhancements to its C4ISR (command, control, communications, computer, intelligence, surveillance, and reconnaissance) systems. These C4ISR systems will also help improve DOD's joint lethality, to help the force maneuver more effectively, and to help ensure greater resiliency to threats and redundancy to not have one severed link end with a catastrophic impact on mission execution.

Finally, as DOD will continue to work to support the NSS, it sees the value of partnerships (NDS 2018). It will integrate with interagency partners at home while it hopes to leverage coalitions like NATO (NDS 2018, 5).

References

CIA. 2019. *The World Factbook, United States.* https://www.cia.gov/library/publications/resources/the-world-factbook/geos/us.html. Accessed 9 June 2019.

DNI. 2019. *Office of the Director of National Intelligence.* https://www.dni.gov/index.php/what-we-do/members-of-the-ic. Accessed 21 Aug 2019.

INSSG. 2021. Interim National Security Strategic Guidance. https://www.whitehouse.gov/wp-content/uploads/2021/03/NSC-1v2.pdf. Accessed 17 Aug 2021.

JSP. 2018. *Joint Strategic Plan FY 2018–2022 U.S. Department of State and U.S. Agency for International Development.* https://www.state.gov/documents/organization/277156.pdf. Accessed 17 Apr 2018.

Lowenthal, Mark M. 2017. *Intelligence From Secrets to Policy,* 7th ed. Sage Press.

NATO. 2018. *The Secretary General's Annual Report 2018.* https://www.nato.int/cps/en/natohq/topics_164559.htm. Accessed 11 June 2019.

NDS. 2018. *National Defense Strategy.* https://www.defense.gov/Portals/1/Documents/pubs/2018-National-Defense-Strategy-Summary.pdf. Accessed 20 Jan 2018.

NSS. 2017. *National Security Strategy of the United States of America*. http://nss archive.us/wp-content/uploads/2017/12/2017.pdf. Accessed 23 Dec 2017.

Weaver, John M. 2018. The 2017 National Security Strategy of the United States. *Journal of Strategic Security* 11 (1): 62–71.

Weaver, John M., and Benjamin Johnson. 2020. *Cyber Security Challenges Confronting Canada and the United States*. New York, USA: Peter Lang Publishing.

Wright, David. 2009. *North Korea's Missile Program*. https://www.ucsusa.org/sites/default/files/legacy/assets/documents/nwgs/north-koreas-missile-pro gram.pdf. Accessed 10 Jan 2018.

CHAPTER 4

Al Qaeda

Abstract Al Qaeda is a transnational terror organization. As a threat, this chapter explores how al Qaeda is using cyber operations to weaken the United States.

Keywords Al Qaeda · Cyber · Communications

BACKGROUND

The most recent former Director of the Central Intelligence Agency squarely stated in a speech in May 2019, that al Qaeda remains a central terror actor and poses a significant threat for the United States (Haspel 2019). The exploitation of cyber TTPs by al Qaeda and similar terrorist organizations are often grouped within three general uses: recruitment and the spread of ideology, the coordination and facilitation of operations and attacks, and lastly for economic gains in progression to fund operations and attacks. None are mutually exclusive as they can overlap and be pursued simultaneously. Rudner made the argument that al Qaeda straddles a continuum between state actors (such as Russia and China) and non-state actors (like hacktivists) and, thus, cyber-terrorism has become a critical strategy for the organization, especially after the death of Osama bin Laden (2013, 455). Likewise, Platt writes that al Qaeda members

J. M. Weaver, *The U.S. Cybersecurity and Intelligence Analysis*, https://doi.org/10.1007/978-3-030-95841-1_4

are routinely frequenting 'how-to' websites to learn hacking techniques (2012, 158).

Al Qaeda traces its lineage to the 1980s in Afghanistan through a network of Arab volunteers who united to more aptly respond to the Soviet invasion with Osama Bin Laden forming a key part of the group's leadership. It has subsequently grown to become a global organization. The guiding mandate of al Qaeda is to unite the worldwide Muslim community to unseat governments that are perceived as un-Islamic; ultimately, it wants to establish a pan-Islamic caliphate under a strict Salafi Muslim interpretation of sharia. Even after the recent targeted killing of the al Qaeda leader in the Arabian Peninsula (al-Rimi), the organization is resilient and still remains a threat.

The following analysis looks et al. Qaeda and how it is making use of cyber to enhance its use of and by targeting the diplomacy, information, military, and economic resources of the United States through the YIRTM-C (M) to weaken this country. More specific evidence from the data is reflected in Annex 4.1.

ANALYSIS

Al Qaeda has helped ensure its survival by minimizing its detection using alternate internet platforms like the 'dark web' to share links (CSIS Report 2018, 20). As this terror organization looks to expand its influence, al Qaeda is implementing encryption techniques to safeguard its content, making it increasingly difficult for intelligence professionals to definitively identify the true intentions of this organization (CSIS Report 2018, 20).

It uses a multitude of social media accounts to influence people abroad and identify potential followers' locations throughout the planet (NTFRA 2018, 12). It does so to expand its base and subsequent global reach; it includes such groups as the Algerian-based Salafist Group for Preaching and Combat (GSPC) (Mendelsohn 2016, 2). More to the point, the merger was announced using a multitude of online media venues (Al-Sahab 2006). Online venues have repeatedly been used to burgeon membership to its ranks (Droukdal 2008). Al Qaeda also uses the internet to promote its *Globalization of Martyrdom* (Moghadam 2008). At times though, its messaging has upset members of the ranks leading to many of them severing ties with the terror organization (Filiu 2009). Silber (2012, 292) provided amplification about al Qaeda's communications process when he considered 16 case studies involving the terror organization, several of which looked at targets in the United States. Though successful implementation of coordinated attacks appears to have waned because of allied counter-terrorism efforts, communications taking place among cells and to leaders regarding attacks focusing on both economic and military targets were fostered by information transference using a variety of cyber conveyances to communicate (Silber 2012, 292–293).

When turning to the United States, the most notable plan since the 9/11 attacks involved al Qaeda implementing 'Operation High Rise'—a plan that focused on deploying and detonating explosive devices in New York City (Silber 2012, 153). Al Qaeda used email as a TTP to communicate with what they called their 'link man' to bring the plot to fruition (Silber 2012, 163).

Al Qaeda adapts and has done so effectively in the wake of coalition efforts by NATO (and by extension, the United States) and is quite adept at using the internet to not only spread its ideology, but to personalize its messaging to expand its ranks (Baken and Mantzikos 2015, 97–117). As a result, al Qaeda and its broad, decentralized network of followers have managed to survive and has remained as a vibrant terror organization even after almost two decades of war with NATO. It remains a threat even today to western nations (including the United States) and remains resolutely committed to spreading its narrative (Dunford 2017). This terror organization sees cyber, messaging, the internet, and email as ways to direct and proactively move operatives into position, identify and target organizations, individuals, and more (Springer 2017, 6–7). That stated, it is not as effective as the Islamic State; more will follow on this in a later chapter (Springer 2017, 6). The threat remains, and al Qaeda is steadfastly committed to using CNO to launch CNE and CNA efforts to harm the United States and its interests, and its allies; cyber-attacks are often far harder to detect/trace back to the source and to prevent than physical attacks employed by conventional terrorists (Springer 2017, 325). What is more surprising, is al Qaeda's skill and competency in making use of the web to raise funds, proselytize, and inspire others to launch attacks (Springer 2017, 346).

Moreover, many transnational non-state and state actors are using cyber as a TTP to enhance their economic positions (Weaver and Johnson 2020). Accordingly, al Qaeda knows this and as Clarke (2018, XXVII) points out, it could look to disrupt the banking sector, foster alternative markets to generate income, engage in theft, and conduct transactions secretly and instantaneously.

Al Qaeda sees the utility of the web in its ability to serve as a pathway and tool to promote its ideology efficiently and without the need for significant monetary resources (Burnett 2019). Arguably, this has been the most prolific aspect of al Qaeda's cyber abilities (Weaver and Johnson 2020). Moreover, al Qaeda has been able to generate income through social media (and, more pointedly, through online payment systems and banking) (NTFRA 2018, 12). It has seen success at also generating money from within the United States through charities that al Qaeda has set up, and even generated money through low-level criminal activity (NTFRA 2018, 12). Likewise, it has made use of banking to gain income from holding hostages for ransom (Callimachi 2014). The growth of cryptocurrencies and blockchain technology (which are largely unregulated and offer several advantages over traditional banking for those with a desire to avoid surveillance detection) creates a layer of potential risk as al Qaeda (and other terrorist organizations) uses them to support their activities (Dion-Schwarz et al. 2019).

Al Qaeda, as a non-state actor, understands the power that the internet provides in its execution of asymmetric offensives against nation-states. More pointedly, this group has been able to use the internet to rebuild and regroup its militant capabilities (Burnett 2019). It, like many other actors, is committed to engaging in activities oriented around causing mass disruption (GAO-19-240SP 2018, 3).

Moreover, al Qaeda has shown its proclivity in using the internet to garner skills in CNO more broadly, and CNA more specifically (Weaver and Johnson 2020). Al Qaeda constantly is looking to develop web hacking tools and even turns to programming to better focus these instruments. Though not seen as kinetically damaging as one sees with attacks from direct fire weapons or area damage done by bombs and explosives, al Qaeda has shown its passion to move from bombs to bytes to exact damage on targets that it chooses (to include infrastructure) from a distance (Weaver and Johnson 2020).

Cyber is an attractive way for al Qaeda to lend support to its operations because developing offensive cyber capabilities is often far less costly than investing in traditional foot soldiers and weapon systems (Weaver and Johnson 2020). Al Qaeda, like many state actors, appears to increasingly understand the cyber domain as a way to provide itself with a stand-off capability keeping its operatives out of harms' way. Al Qaeda, by exploiting cyber, can exert great damage less expensively than by pursuing its traditional TTPs of the past. Moving beyond this, one can see the possibility of conducting multiple attacks simultaneously, thereby creating more confusion and damage to its targets (which could include those in the United States).

When referring to the YIRTM-C (M) and the evidence in Annex 4.1, several points of interest emerge. When turning to the instruments of power found in the model in Fig. 2.1, al Qaeda overwhelming is leveraging the information component of the model; it uses this to draw members into its ranks mostly by using social media and as a way to use the web to generate money. Two more components are used by al Qaeda, but to a lesser extent than information and these include the military and economic components; more specifically, efforts included growing its members and to move money in step with the information instrument from the model. This terror organization was not that effective at using diplomacy to draw other organizations to support its efforts, nor was it adept at negating the diplomatic efforts of the United States.

ANNEX 4.1: AL QAEDA

How/why	D.I.M.E	Source type	Author	Date	Page(s)
The Director of the CIA stated during a speech in May 2019, that al Qaeda and the threat that it poses remains as a central non-state actor for this agency	I, M, E	Speech	Haspel	2019	NP
There is utility in how al Qaeda is able to spread its ideology through cyber	I	Press Release	Burnett	2019	NP
Al Qaeda is an organization that sees the power that the internet provides; it uses the internet to regroup and rebuild its militant capabilities	M	Press Release	Burnett	2019	NP

(continued)

(continued)

How/why	D.I.M.E	Source type	Author	Date	Page(s)
Al Qaeda is bent on engaging in activities oriented around causing mass disruption	I	Government Report	GAO-19-204SP	2018	3
Al Qaeda has learned to adapt in the wake of coalition efforts by NATO (and by extension the United States) and is using the internet more effectively	I, M	Book	Baken and Mantzikos	2015	97–117
Al Qaeda has realized the economic value of cyber to foster alternative markets, disrupt the banking sector, engage in theft, and to conduct transactions quickly and secretively	I, E	Book	Clarke	2018	XXVII
Cyber is a way to readily move about, identify, and target individuals, organizations, and more	I, M, E	Book	Springer	2017	6–7
The threat remains, and al Qaeda is committed to using CNO as a way to launch CNA and CNE efforts to harm the United States (and its allies)	I, M, E	Book	Springer	2017	325
Al Qaeda is competent in using the web to fundraise, proselytize, and inspire others to launch attacks	I, M	Book	Springer	2017	346
Al Qaeda made use of such information technology as cell phones, beepers, the internet, and memory sticks to plan attacks	I, M	Book	Silber	2012	258

(continued)

(continued)

How/why	D.I.M.E	Source type	Author	Date	Page(s)
Communications among cells and to leaders regarding attacks focus on both military and economic targets; these were fostered by information transference using a variety of means to communicate	I, M, E	Book	Silber	2012	258–259
Al Qaeda was using email as a way to communicate with key members	I, M	Book	Silber	2012	163
Al Qaeda uses a multitude of social media accounts to influence people abroad and identify potential followers' locations around the world	I	Government Report	NTFRA	2018	12
Al Qaeda raises money through social media (and through internet banking and online payment systems) to generate money	I, E	Government Report	NTFRA	2018	12
Al Qaeda has been successful at also generating money from the United States through charities that it has set up and through low-level criminal activity	I, E	Government Report	NTFRA	2018	12
Al Qaeda remains a threat to western countries (to include the United States and allies alike)	I	Speech	Dunford	2017	NP
It uses cyber to expand its base to include such groups as the Algerian-based Salafist Group for Preaching and Combat (GSPC)	I	Book	Mendelsohn	2016	2

(continued)

(continued)

How/why	D.I.M.E	Source type	Author	Date	Page(s)
Al Qaeda's messaging, at times, has disenfranchised members causing them to leave al Qaeda	I	Journal	Filiu	2010	217–220
Online venues have increased membership to its ranks	I	Interview	Droukdal	2008	NP
Al Qaeda uses the internet to foster its *Globalization of Martyrdom*	I, M	Book	Moghadam	2008	61–62

References

Al-Sahab. 2006. Hot Issues Interview with Shaykh Ayman al-Zawahiri. *OSC*.
Baken, Denise N., and Ioannis Mantzikos. 2015. *Al Qaeda: The Transformation of Terrorism in the Middle East and North Africa*. Santa Barbara, CA: Praeger.
Burnett, Michael. 2019. Winning the War Against Terror Messaging. https://www.whitehouse.gov/articles/winning-war-terrorist-messaging/. Accessed 22 July 2019.
Callimachi, Rukmini. 2014. Paying Ransoms, Europe Bankrolls Qaeda Terror. *New York Times*.
Clarke, Colin P. 2018. *Terrorism: The Essential Reference Guide*. ABC-CLIO, LLC.
CSIS Report. 2018. *2018 CSIS Public Report*. https://www.canada.ca/content/dam/csis-scrs/documents/publications/2018-PUBLIC_REPORT_ENGLISH_Digital.pdf. Accessed 21 Aug 2019.
Dion-Schwarz, C., D. Manheim, and P. B. Johnston. 2019. *Terrorist Use of Cryptocurrencies*. Rand Corporation. https://www.rand.org/pubs/research_reports/RR3026.html. Accessed 6 Mar 2020.
Droukdal, Abdelmalek. 2008. *An Interview with Abdelmalek Droukdal*. New York Times.
Dunford, Joe. 2017. Department of Defense Press Briefing by General Dunford in the Pentagon Briefing Room. https://www.defense.gov/Newsroom/Transcripts/Transcript/Article/1351411/department-of-defense-press-briefing-by-general-dunford-in-the-pentagon-briefin/. Accessed 17 Dec 2019.

Filiu, Jean-Pierre. 2009. The Local and Global Jihad of al-Qaeda in the Islamic Maghrib. *Middle East Journal* 63 (2): 217–220.

GAO-19-204SP. 2018. Report to Congressional Committees National Security Long-Range Emerging Threats Facing the United States as Identified by Federal Agencies. https://www.gao.gov/assets/700/695981.pdf. Accessed 23 July 2019.

Haspel, Gina. 2019. CIA Director Gina Haspel Speaks at Auburn University. https://www.cia.gov/news-information/speeches-testimony/2019-speeches-testimony/dcia-haspel-auburn-university-speech.html. Accessed 17 June 2019.

Mendelsohn, Barck. 2016. *The al-Qaeda Franchise: The Expansion of al-Qaeda and Its Consequence*. Oxford Scholarship Online.

Moghadam, Assaf. 2008. The Globalization of Martyrdom. Al Qaeda, Salafi Jihad, and the Diffusion of Suicide Attacks. *Johns Hopkins University Press*.

NTFRA. 2018. National Terrorism Financing Risk Assessment. https://home.treasury.gov/system/files/136/2018ntfra_12182018.pdf. Accessed 17 Dec 2019.

Platt, V. 2012. Still the Fire-Proof House? An Analysis of Canada's Cybersecurity Strategy. *International Journal* 67 (1): 155–167.

Rudner, M. 2013. Cyber-Threats to Critical National Infrastructure: An Intelligence Challenge. *International Journal of Intelligence and CounterIntelligence* 26: 453–481.

Silber, Mitchell D. 2012. *The Al Qaeda Factor*. Philadelphia: University of Pennsylvania Press.

Springer, Paul J. 2017. *Encyclopedia of Cyber Warfare*. ABC-CLIO, LLC.

Weaver, John M., and Benjamin Johnson. 2020. *Cyber Challenges Confronting Canada and the United States*. New York, USA: Peter Lang Publishing.

CHAPTER 5

People's Republic of China (China)

Abstract China is seen as the most prominent threat to the United States. This chapter affords consideration to how it is using cyber operations to weaken the United States.

Keywords China · Xi Jinping · PLA

BACKGROUND

The People's Republic of China (PRC) has 1.3 billion citizens placing it in the position as the largest in the world in terms of national populations (CIA 2019). China's capital is Beijing and in 1949 the PRC emerged after the rise of Mao Zedong, Chairman of the Communist Party of China, assumed the role as its leader (CIA 2019). The country's legal system is influenced by both continental European and Soviet civil law systems (CIA 2019).

When shifting to its political system and under the executive branch, the head of government is President Xi Jinping; he has remained in office since March of 2013 (CIA 2019). The president is indirectly elected by the National People's Congress for a five-year term with no limits on the number of terms that one can serve (CIA 2019). The cabinet is a

© The Author(s), under exclusive license to Springer Nature Switzerland AG 2022
J. M. Weaver, *The U.S. Cybersecurity and Intelligence Analysis*,
https://doi.org/10.1007/978-3-030-95841-1_5

State Council; it is appointed under the National People's Congress (CIA 2019).

Under the legislative branch, China's is unicameral; it is the National People's Congress (CIA 2019). The members are elected indirectly by congresses at the municipal, regional, and provincial levels (CIA 2019).

China's judicial branch consists of the Supreme People's Court (at the highest level); it includes a chief justice, 13 grand justices, and there are over 340 judges in total (CIA 2019). Those in the ranks of chief justices are appointed by the People's National Congress for five-year terms; they can serve no more than two consecutive terms (CIA 2019).

China has several subordinate courts. These include the Higher People's Court, the Intermediate People's Court, District and County People's Courts, Autonomous Region People's Court, Special People's Courts for military, maritime, transportation, and forestry issues, and lastly, the International Commercial Courts (CIA 2019).

Economically, China has slowly moved from a closed, centrally planned system since the late 1970s to one where its economy is more market-oriented but where the state continues to have a central role in the management of its markets as well as the internationalization of Chinese capital in terms of foreign investment (CIA 2019). In recent decades, China has incrementally implemented reforms; it has resulted in efficiency gains that have led to an increase in GDP since 1978 by a tenfold factor (CIA 2019). Its reforms began with the phaseout of collectivized agriculture and emerged to now include increased autonomy for state enterprises, the gradual liberalization of prices, the growth of the private sector, fiscal decentralization, development of a modern banking system, and stock markets that are opening to foreign trade and investment (CIA 2019). China continues its pursuit of an industrial policy, has a restrictive investment regimen, and the government provides state support of key sectors (CIA 2019). China also had one of the quickest-growing economies from 2013 to 2017 on the planet, which averaged slightly greater than 7% real growth per year (CIA 2019). When looking at the purchasing power parity (PPP) basis and after adjusting for price differences, China rose to the world's largest economy in 2017, surpassing the United States for the first time in modern history (CIA 2019). China has also emerged as the largest exporter in the world in 2010, and then emerged as having the largest trading nation status in 2013 (CIA 2019). Conversely, China's per capita income is well below the average in the world (CIA 2019).

When turning to China's military, it is composed of a diverse military force capable of conducting operations on the air, land, and sea. It is made up of the People's Liberation Army (PLA), the Navy (PLAN), which includes marines and naval aviation, an Air Force (PLAAF) that includes airborne forces, a Rocket Force (which serves as its strategic missile force), Strategic Support Force (China's space and cyber forces), the People's Armed Police (PAP), which includes its Coast Guard, and finally, the PLA Reserve Force (CIA 2019).

The subsequent analysis looks at China and how it is using cyber through its targeting of the diplomatic efforts, information, military, and economic resources of the United States while enhancing its components of the D.I.M.E. through the YIRTM-C (M). One will find more specific evidence drawn from the data in Annex 5.1.

ANALYSIS

Biden sees China as a significant rival (INSSG 2021, 6). China's deep economic interdependence with the United States makes cyber a logical avenue for strategic development by the PRC as any blatant offensive conventional measure would most likely damage those relationships. Likewise, China has and will most likely continue to invest in cyber to influence elections (WTA 2021). In contrast, cyber helps China achieve an element of deniability through 'grey zone' operations (Weaver and Johnson 2020). In turn, this presents a unique situation for the United States as it becomes increasingly apparent for the need of this country to balance national security concerns with the needs of economic development. Indeed, balancing these concerns is evidenced by the ongoing issue regarding whether China should be allowed access to telecommunication networks through Chinese multinationals (for example—Huawei's desire to help construct other allies' 5G networks). There is growing evidence to suggest that said involvement may undermine national security (see Lombardi 2020).

Additionally, the recent charges asserted by the U.S Justice Department against four Chinese military members for the massive 2017 Equifax hack [this affected approximately 145 million Americans and it compromised their personal identifiable information (PII), including social security numbers] indicates that Beijing is significantly leveraging cyber to undermine national security and societal security more widely (CBC 2020). Generally speaking, the PRC has used cyber TTPs to give it an edge over nations with more advanced militaries (Weaver 2017). China, along with other nation-states, has been committed to using cyber to potentially cause mass disruption (GAO-19-204SP 2018, 3). Rudner (for example) notes that China's "People's Liberation Army (PLA) is reported to have deployed a dedicated signals intelligence unit for cyber-espionage" (2013, 454). China has also been found complicit in penetrating western defense systems and it has gained access to sensitive military documents of its neighbors in the Pacific region (Platt 2012, 160).

Likewise, the U.S information infrastructure and cyberspace networks are vulnerable and thus have become an easy target by state actors to include China (Pomeroy 2019). At present, cyber activities by potential adversaries (including China) have led to an erosion of the U.S. military advantages; this has adversely affected the United States' economic prosperity (Pomeroy 2019). Moreover, China also engaged in cyber espionage to aid in the compression of the research and development timelines, especially when seeking to acquire information on technology more broadly and specifically, in its pursuit of advanced weapon designs. Under China's military, it has a strategic support force that includes space and cyber forces (Weaver 2019). China has shown proficiency in using cyber, and by extension, information warfare on the global stage (Rogers 2017) and is using cyber as a conduit to limit western influences around the planet, but especially U.S influence in the Indo-Pacific region (NCS 2018, 2).

China's intelligence service is quite extensive and adept at conducting both espionage and cyber activity (Lowenthal 2017, 499). Lowenthal (2017, 499) states that this intelligence service has two main purposes: (1) to direct activities against identified dissidents (for internal security reasons) and (2) to conduct intelligence collection/operations abroad. The conduct of these operations falls under the direction of the Ministry of State Security though it doesn't compare to the power possessed by China's Central Military Commission; often the two are at odds with one another (Lowenthal 2017).

A recent U.S. Defense Intelligence Agency (DIA) press release shows that China is investing in several types of activities (diplomatically, militarily, and economically) to enhance its industrial base and infrastructure (DIA Press Release 2019). More pointedly, the PRC's Army (PLA) sees information dominance as a necessity to achieving victory on the modern battlefield (China Military Power 2019). China is also committed to conducting cyber reconnaissance, computer network attack (CNA), and computer network defense (CND) to enhance its position while it looks to improve command and control of its armed forces under its Strategic Support Force (SSF) (China Military Power 2019).

Moreover, and dating back to the founding of the SSF in December of 2015, China has formed unique cyber elements under the PLA's General Staff Third (specializing in technical reconnaissance), and Fourth (specializing in electronic countermeasure and radar) Departments (China Military Power 2019). The PRC has done so because it sees the utility of cyberspace as a new pillar to help it achieve economic prosperity, and it sees the inextricable linkage of cyberspace to the growth of its economy (Chinese Strategy 2015). China is looking to expedite the development of a cyber force as cyberspace continues to weigh more in military security. China wants to improve its cyberspace situational awareness capabilities, cyber defense, support for this country's endeavors in cyberspace, and finally, participation in international cyber cooperation (Weaver and Johnson 2020). It is doing so to ensure national network and information security, to minimize the likelihood of a major cyber crisis, and maintain its national security and social stability (Chinese Strategy 2015). China is very concerned about its cyber network defense (CND) to protect itself more aptly from exploitation by others (state and non-state actors) that might engage in similar behavior (Weaver 2019).

China, accordingly, could then use cyber reconnaissance to enable the PLA to collect technical and operational data to be used as intelligence while it assesses the TTPs to look for vulnerabilities to launch CNA. Likewise, Chinese cyber forces could theoretically target command and control (C2) nodes like those of NORAD and command, control, communications, computers, intelligence, surveillance, and reconnaissance (C4ISR) assets, as well as logistics' centers of gravity (China Military Power 2019).

As part of its overall defense plan, the PRC will use cyber to implement denial and deception (D2) operations. It will most likely do so to help inhibit the effectiveness of adversarial powers' intelligence services to gain granularity into understanding China's real intentions (China Military Power 2019). More specifically, it could employ counter reconnaissance measures to evade, jam, and/or destroy the full spectrum of efforts employed by enemy intelligence to acquire information on PLA units, facilities, and its leadership (Weaver and Johnson 2020).

China operates both regional and global satellites for a variety of reasons (Security in Space 2019, 8). More specifically, it has made the investment in space to better leverage information technology and cyber for both diplomatic and economic gains (Security in Space 2019, 13).

Of all the actors studied in this book, the PRC is investing the most in space-based systems to counter what its leaders perceive as an over-reliance by western nations on technology (SFTR1 2018; SFTR2 2018). Thus, China's PLA will look to use space as a way to conduct information warfare (Annual Report to Congress 2017).

The Pentagon has warned that China is implementing a moderniza-tion and development strategy that includes proprietary theft (among other strategies) for the acquisition of foreign technologies (both military and dual use) to shorten the research and development (R&D) timeline (Department of Defense 2019, iii). China realizes that instead of investing in its R&D programs to create technological innovations, that it is far less expensive and less time-consuming if they steal it. During a speech given to The New America Foundation in 2017, when discussion ensued with Admiral Rogers [the former head of the U.S. National Security Agency (NSA)], a statement was made that China was singled out as a frequent culprit in stealing secrets vis-à-vis cyber (both government and private) from the United States (Rogers 2017).

The PRC will most likely continue to conduct cyber espionage to illegally obtain information from private sector and government organi-zations (WTA 2019, 5). Moreover, China hopes to further compress the R&D timeline to reduce the expenditures and time that it would take to get comparable products to production to compete with those of western nations (Weaver 2017; Weaver and Johnson 2020).

However, there are realities associated with limits to achieving techno-logical parity with the United States through time-compression strategies and one must appreciate this, especially concerning military technologies given the increasing complexity and costs, which can create barriers to entry for the development and production of advanced weapon systems (Gilli and Gilli 2019, 142). Nonetheless, this may further incentivize this reliance on cyber platforms as these potentially represent a relatively cost-efficient and practical way of achieving a level of technological and power symmetry with more advanced capabilities and technologies of the United States and other allies.

The PRC also uses its intelligence and security agencies for cyber espi-onage on firms of the United States and its allies (WTA 2019, 5). The United Kingdom (UK) formally asserted that the Chinese Ministry of

State Security was being referred to as "APT 31" and "APT 40" (UK 2021). The UK's National Cyber Security Centre (NCSC) assessed that it was almost certain that APT 31 has targeted political figures and government entities (UK 2021). The NCSC also believes that it is highly likely that the Chinese, through APT 40 have targeted naval defense contractors and the maritime industries of the United States and other European countries (UK 2021). Likewise, China has the intent to scope out vulnerabilities of critical infrastructure to exploit these at a time and place of their choosing should it feel threatened (WTA 2019). It has done so in the past; a specific incident involved China causing disruption of a natural gas pipeline in the United States. There have also been links to attacks on the U.S power grid (WTA 2019).

Other assessments of China's People's Liberation Army capabilities also show the relevancy of this country's acquisition of capabilities and systems for the intended use for cyber-attacks (Regional Focus 2015). Accordingly, it sees cyber as a complementary TTP for it to realize gains through asymmetry to engage in the stealing of industry and military secrets, espionage, and more according to the U.S. Assistant Secretary of Defense for Indo-Pacific security affairs (Schriver 2019). As China continues its global burgeoning, it is marshaling diplomatic, economic, and military resources through cyber TTPs to foster its rise as both a regional and global player (GAO-19-204SP 2018, i).

Economically, China has been actively pursuing cyber espionage in illegally acquiring proprietary information (Weaver and Johnson 2020). Moreover, it has been engaged in the hacking of Microsoft networks throughout the world to acquire intellectual property (UK 2021). What's more, in a White House press release in July 2021, the Biden Administration asserted that China has been complicit in cyber enabled extortion, ransomware attacks, crypto-jacking, and more for its financial gain (White House 2021). It has done so, as illustrated above, to compress the research and development timeline to enhance the PRC's position in the world while improving its military capabilities (Weaver 2017; Weaver and Johnson 2020).

When returning to the YIRTM-C (M), China has relied heavily on three of the instruments to help advance its position; it has done so while also hurting the position of the United States on the world stage (Weaver and Johnson 2020). As was the case with al Qaeda, and as demonstrated through the evidence found in Annex 5.1, China, overwhelmingly, has made use of the information component of the model. The PRC has demonstrated its adeptness at weakness exploitation of industrialized countries to gain access to proprietary information to reverse engineer the production of both weapon systems and commercially available goods (Weaver and Johnson 2020).

By extension, China has used cyber to help improve its military capabilities, using cyber units to expose and exploit weaknesses in weapon systems. This has especially been the case with those that are linked vis-à-vis technology; it has used its prowess in cyber to gain economic advantages to grow its economy at a rate faster than that of the United States (Weaver and Johnson 2020).

Though China has been actively fomenting diplomatic relations with other countries (especially throughout Asia and Africa), it has not used cyber effectively to do so as Russia has regarding western elections (Weaver and Johnson 2020). Likewise, China realizes the many advantages that cyber provides in terms of asymmetric warfare; it will most likely not want to formalize restrictions too greatly in limiting what it can and cannot do in cyberspace (Weaver and Johnson 2020).

ANNEX 5.1: CHINA

How/why	D.I.M.E	Source type	Author	Date	Page(s)
The DIA shows that China is investing in all types of projects (diplomatically, militarily, and economically) to enhance its infrastructure and industrial base	D, M, E	Press Release	DIA Press Release	2019	NA
The PLA sees information dominance as a precursor to achieving victory	I	Government Document	China Military Power	2019	45
China leans on cyber reconnaissance, CNA, and CND to enhance its position while looking to improve command and control of its armed forces	I, M	Government Document	China Military Power	2019	45

(continued)

(continued)

How/why	D.I.M.E	Source type	Author	Date	Page(s)
China has created specific cyber elements which includes a technical reconnaissance and an electronic countermeasure and radar sections	I, M	Government Document	China Military Power	2019	45
China's cyberspace is a key pillar for its economic prosperity	I, E	Plans and Systems	Chinese Strategy	2015	NA
China wants to safeguard its security and interests in new domains	I	Plans and Systems	Chinese Strategy	2015	NA
China looks to expedite the development of a cyber force, and improve its capabilities of cyberspace situational awareness, and cyber defense, in order to support this country's endeavors in cyberspace	I	Plans and Systems	Chinese Strategy	2015	NA
Chinese cyber forces could target command and control (C2) nodes like those of NORAD and command, control, communications, computers, intelligence, surveillance, and reconnaissance (C4ISR) assets, as well as logistics centers of gravity	I, M	Government Document	China Military Power	2019	46
Cyber is a way to engage in D2 operations	I, M	Government Document	China Military Power	2019	46
China is investing the most in space-based systems to counter western nations	I, M, E	Testimony	SFTR1, SFTR2	2018	NA
It has invested in space, to leverage information technology and cyber for both diplomatic and economic purposes	D, E	Government Document	Security in Space	2019	13
China's PLA looks to use space as a way to conduct information warfare	I, M	Government Document	Annual Report to Congress	2017	39

(continued)

(continued)

How/why	D.I.M.E	Source type	Author	Date	Page(s)
The PRC will most likely continue its conduct of cyber espionage to illegally obtain information from government and private sector organizations	I, E	Government Document	WTA	2018	5
China has the intent to seek weaknesses of critical infrastructure to exploit at a time and place of their choosing should they feel threatened	I, E	Government Document	WTA	2018	5
The PRC exploits cyber TTPs in order to give it an edge over countries with more advanced militaries	I, M	Book Chapter	Weaver	2017	29–54
Under its military, China has a strategic support force that is comprised of space and cyber forces	I, M	Book	Weaver	2019	29–54
China has an extensive intelligence service skilled at conducting both espionage and cyber activity	I, M	Book	Lowenthal	2017	499
China proficiently uses cyber, and by extension, information warfare, globally	I, M, E	Interview	Rogers	2017	NA
Admiral Rogers stated that China was a frequent culprit in stealing secrets vis-à-vis cyber (both private and government) from the United States	I, E	Interview	Rogers	2017	NA
U.S. information infrastructure and cyber space networks have been vulnerable in the past and thus becoming an easy target by China	I, M, E	Book Chapter	Pomeroy	2019	315–225

(continued)

(continued)

How/why	D.I.M.E	Source type	Author	Date	Page(s)
Cyber activities by the U.S. adversaries have eroded the U.S. military advantages and adversely impacted the U.S. economic prosperity	M, E	Book Chapter	Pomeroy	2019	315–325
China is using cyber to harm other countries' influence in the world	D	Government Document	NCS	2018	2
China is committed to using cyber to potentially cause mass disruption	I	Government Report	GAO-19-204SP	2018	3
Biden sees China as a significant rival	D, M, E	Government Report	INSSG	2021	6
China has and will most likely continue to invest in cyber as a way to influence elections	D	Government Report	WTA	2021	8
The UK's National Cyber Security Centre (NCSC) assessed that it was almost certain that APT 31 has targeted political figures and government entities	D	Government Report	UK	2021	NP
The NCSC also believes that it was highly likely that the Chinese, through APT 40 have targeted naval defense contractors and the maritime industries of the United States and other European countries	M, E	Government Report	UK	2021	NP

REFERENCES

Annual Report to Congress. 2017. *Military and Security Developments Involving the People's Republic of China 2017*. Office of the Secretary of Defense.

CBC. 2020. U.S. Accuses Chinese Military Hackers in Massive Equifax Breach Over 2 Years Ago. https://www.cbc.ca/news/business/us-justice-charges-china-equifax-1.5458110.

China Military Power. 2019. *China Military Power: Modernizing a Force to Fight and Win*. https://www.dia.mil/Portals/27/Documents/News/Mil itary%20Power%20Publications/China_Military_Power_FINAL_5MB_201 90103.pdf. Accessed 26 Feb 2020.

Chinese Strategy. 2015. *China's Military Strategy*. https://news.usni.org/2015/ 05/26/document-chinas-military-strategy. Accessed 10 June 2019.

CIA. 2019. The World Factbook, China. https://www.cia.gov/library/publicati ons/resources/the-world-factbook/geos/ch.html. Accessed 17 June 2019.

Department of Defense. 2019. *Annual Report to Congress: Military and Security Developments Involving the People's Republic of China 2019*. Office of the Secretary of Defense. https://media.defense.gov/2019/May/02/ 2002127082/-1/-1/1/2019_CHINA_MILITARY_POWER_REPORT.pdf. Accessed 6 Mar 2020.

DIA Press Release. 2019. *Defense Intelligence Agency Releases Report on China Military Power*. https://www.jcs.mil/Media/News/News-Display/ Article/1732847/defense-intelligence-agency-releases-report-on-china-mil itary-power/. Accessed 10 June 2019.

GAO-19-204SP. 2018. Report to Congressional Committees National Security Long-Range Emerging Threats Facing the United States as Identified by Federal Agencies. https://www.gao.gov/assets/700/695981.pdf. Accessed 23 July 2019.

Gilli, A., and M. Gilli. 2019. Why China Has Not Caught Up Yet: Military-Technological Superiority and the Limits of Imitation, Reverse Engineering, and Cyber Espionage. *International Security* 43 (3): 141–189.

INSSG. 2021. Interim National Security Strategic Guidance. https://www.whi tehouse.gov/wp-content/uploads/2021/03/NSC-1v2.pdf. Accessed 17 Aug 2021.

Lombardi, M. 2020. Why Ottawa Must Say No to Huawei on Building Canada's 5G Networks, January 2.

Lowenthal, Mark M. 2017. *Intelligence from Secrets to Policy*, 7th ed. Sage Press.

NCS. 2018. National Cyber Strategy of the United States of America. https:// www.whitehouse.gov/wp-content/uploads/2018/09/National-Cyber-Str ategy.pdf. Accessed on July 22, 2019 and Prosperity in the Digital Age. https://www.securitepublique.gc.ca/cnt/rsrcs/pblctns/ntnl-cbr-scrt-strtg/ ntnl-cbr-scrt-strtg-en.pdf. Accessed 23 July 2019.

Platt, Victor. 2011–2012. Still the Fire-Proof House? An Analysis of Canada's Cybersecurity Strategy. *International Journal* 67 (1): 155–167.

Pomeroy, Jennifer Yongmei. 2019. Challenges of the U.S. National Security and Moving Forward (Chapter). In *Global Intelligence Priorities (from the Perspective of the United States)*. Nova Science Publishers.

Regional Focus. 2015. *Regional Focus Asia Pacific.* http://www.janes.com/article/39339/regional-focus-asia-pacific-es14e2. Accessed 18 Feb 2016, 17 June 2014.

Rogers, Michael. 2017. Admiral Michael S. Rogers (USN), Director, National Security Agency, and Commander, U.S. Cyber Command, Delivers Remarks at The New America Foundation Conference on CYBERSECURITY. https://www.nsa.gov/news-features/speeches-testimonies/speeches/022315-new-america-foundation.shtml. Accessed 20 July 2017.

Rudner, Martin. 2013. Cyber-Threats to Critical National Infrastructure: An Intelligence Challenge. *International Journal of Intelligence and Counterintelligence* 26 (3): 453–481.

Schriver, Randall G. 2019. DOD Official Details Continuing Chinese Military Buildup. https://dod.defense.gov/News/Article/Article/1836512/dod-official-details-continuing-chinese-military-buildup/. Accessed 21 July 2019.

Security in Space. 2019. *Challenges to Security in Space.* https://www.dia.mil/Portals/27/Documents/News/Military%20Power%20Publications/Space_Threat_V14_020119_sm.pdf. Accessed 10 June 2019.

SFTR1. 2018. Statement for the Record: Worldwide Threat Assessment of the U.S. Intelligence Community. Office of the Director of National Intelligence, 13 Feb 2018.

SFTR2. 2018. Statement for the Record: Worldwide Threat Assessment of the U.S. Intelligence Community. Office of the Director of National Intelligence, 13 Feb 2018.

UK. 2021. UK and Allies Hold Chinese State Responsible for a Pervasive Pattern of Hacking. https://www.gov.uk/government/news/uk-and-allies-hold-chinese-state-responsible-for-a-pervasive-pattern-of-hacking. Accessed 20 July 2021.

Weaver, John M. 2017. Cyber Threats to the National Security of the United States: A Qualitative Assessment (Chapter). In *Focus on Terrorism*, vol. 15. Nova Science Publishers.

Weaver, John M. 2019. *United Nations Security Council Permanent Member Perspectives Implications for U.S. and Global Intelligence Professionals.* Peter Lang Publishing.

Weaver, John M., and Benjamin Johnson. 2020. *Cyber Security Challenges Confronting Canada and the United States.* New York, USA: Peter Lang Publishing.

White House. 2021. The United States, Joined by Allies and Partners, Attributes Malicious Cyber Activity and Irresponsible State Behavior to the People's Republic of China. https://www.whitehouse.gov/briefing-room/statements-releases/2021/07/19/the-united-states-joined-by-allies-and-partners-attributes-malicious-cyber-activity-and-irresponsible-state-behavior-to-the-peoples-republic-of-china/. Accessed 20 July 2021.

WTA. 2019. *Worldwide Threat Assessment of the US Intelligence Community.* https://completethreatpreparedness.com/wp-content/uploads/2019/02/2019-ODNI-Worldwide-Threat-Assessment.pdf. Accessed 11 June 2019.

WTA. 2021. *Worldwide Threat Assessment of the US Intelligence Community.* https://www.dni.gov/files/ODNI/documents/assessments/ATA-2021-Unclassified-Report.pdf. Accessed 6 July 2021.

CHAPTER 6

Islamic Republic of Iran (Iran)

Abstract When turning to the Middle East, the primary threat to stability in this region (and regional allies of the United States) is Iran. Accordingly, this chapter explores how Iran is using cyber operations to weaken the United States.

Keywords Cyber · Raisi · JCPOA · Espionage

Background

Up until 1935, Iran was known as Persia and since 1979, it has been officially referred to diplomatically as the Islamic Republic of Iran. It is a theocratic republic; its capital is Tehran (CIA 2019). Likewise, Iran's legal system is one based on both religion and secular Islamic Law (CIA 2019).

The government of Iran is grounded through its executive, legislative, and judicial branches. The chief of state is the Supreme Leader Ali Hosseini-Khamenei (since June 1989), and the head of government is President Ebrahim Raisi who has been in power since August of 2021 (under the executive branch) (CIA 2019). This branch is rounded out by its cabinet which is known as the Council of Ministers; they are selected by the president but require legislative approval. It is important to note

J. M. Weaver, *The U.S. Cybersecurity and Intelligence Analysis*, https://doi.org/10.1007/978-3-030-95841-1_6

that the Supreme Leader has limited control over the appointment of several of the ministers (CIA 2019).

When shifting to the legislative branch, Iran is a unicameral Islamic Consultative Assembly (Weaver and Johnson 2020). The Assembly is elected in either single or multi-seat constituencies from a two-round vote (CIA 2019).

Under the judicial branch, the highest court is the Supreme Court. The head of the High Judicial Council appoints the president of the Supreme Court which is a five-member body that includes the chief justice of the Supreme Court, the prosecutor general, and three clergies (CIA 2019). The subordinate courts under the judicial branch include Penal Courts I and II, the Courts of Peace, the Islamic Revolutionary Courts, and the Special Clerical Court (CIA 2019).

When turning to Iran's economy, it is one marked by statist policies, inefficiencies, and one with a strong reliance on oil and gas exports (Weaver and Johnson 2020); Iran also possesses significant agricultural, industrial, and service sectors (CIA 2019). Iran's government directly owns and operates hundreds of organizations, and it indirectly has control over many companies with an affiliation to the country's security forces (CIA 2019). Several factors have an adverse effect on its economy; these include subsidies, corruption, price controls, and a banking system replete with billions of dollars of non-performing loans. Collectively, these factors weigh down the economy and have an undermining effect on the potential for economic growth (CIA 2019). Years of sanctions put in place by the United Nations broadly, and the United States more specifically, have further eroded Iran's economic competitiveness (Weaver and Johnson 2020).

Iran's private sector activity includes the following: small-scale workshops, farming, services, and limited manufacturing (CIA 2019). These are added to mining, medium-scale construction, metalworking, as well as cement production (CIA 2019). Corruption is rampant in Iran (Weaver and Johnson 2020); this has resulted in a significant informal market activity that is flourishing (CIA 2019).

Iran's military is one of the most formidable in the Middle East. It is comprised of the Islamic Republic of Iran Regular Forces (Artesh); these include its Ground Forces (Army), an Air Force (IRIAF), a Navy, the Islamic Revolutionary Guard Corps (otherwise known as the Sepah-e Pasdaran-e Enqelab-e Eslami or IRGC) which also is broken down by Ground Forces, the Khatemolanbia Air Defense Headquarters, an

Aerospace Force, and the Qods Force (special operations) (CIA 2019). It also has three primary military organizations dedicated to cyber (Lewis 2019).

Iran has been complicit with its sponsorship of terrorism (Weaver and Johnson 2020). Accordingly, the Islamic Revolutionary Guard Corps' (IRGC) aim is to safeguard Iran's Islamic Revolution, to more aptly foster the spread of Shia influence to shore up its internal security which includes law enforcement, border control, and suppressing domestic opposition (CIA 2019). It also controls the nation's rockets and missiles, and influences Iran's politics and economy (CIA 2019). Conversely, the Qods Force has the express aim to protect Iran's Islamic Revolution (CIA 2019). It, too, has the desire to spread Shia influence to conduct covert overseas operations (often providing direct support to other terrorist organizations that sometimes include Sunni groups like the Taliban when their goals align), by providing monetary resources for terror operations, logistics support enabling these activities, training and/or weapons to commit terror attacks (either directly or through its proxies), and finally, to recruit/train/equip other foreign Islamic revolutionary groups throughout the Middle East (CIA 2019).

The subsequent analysis looks at Iran and how it is using cyber to enhance its position while it also targets the diplomacy, information, military, and economic resources of the United States through the YIRTM-C (M). One will find more specific evidence drawn from the data in Annex 6.1.

ANALYSIS

Iran has pursued capabilities that are game changing to exert influence against the interests of the United States (INSSG 2021, 8). Though most intelligence professionals believe that Iran had been generally complying with the terms of the Joint Comprehensive Plan of Action (JCPOA) (until late 2019) regarding its nuclear program, it has invested extensively in cyber TTPs to exert influence and power throughout the world. In recent remarks by the former leader of the Central Intelligence Agency (CIA), Gina Haspel spoke of the Iranian threat and how this steadily gained prominence and the attention of the agency in recent years (Haspel 2019). Iran previously used cyber to conduct denial of service (DoS) and other attacks against U.S. interests; these included the banking sector and other business operations (Daigle 2020). More recently, the killing of

Iran's military commander, Major General Qassem Soleimani by a U.S. drone strike in Iraq, has led intelligence professionals to warn that Iran will likely use cyber-attacks as a means of retaliation (Daigle 2020; Weaver and Johnson 2020).

Iran has also engaged in disinformation campaigns (CISA Iran 2021a). It is promoting a pro-Iranian message while it looks to foster anti-U.S. sentiment (CISA Iran 2021a). Iran was also actively attempting to use cyber to alter the U.S. presidential election in 2021 (CISA Iran 2021b). It most likely will engage in disinformation campaigns to discredit the United States and to circulate content counter to official U.S. messaging (WTA 2021, 14).

That stated, the Department of Homeland Security in the United States has stated its concerns about increases in Iran's malicious cyber forays directed at this country (Leithauser 2019). Krebs (2019) wrote that there is a pressing need for improvement in cybersecurity and awareness at all levels in the private sector because of vulnerabilities that have been exposed in recent years. Moreover, concerns arise about Iran's potential attacks on the North American telecommunications infrastructure (Leithauser 2020). It has remained committed to disruptive and destructive cyber operations targeting strategic objectives, the financial sector, and energy; its cyber espionage also targets intellectual property (CISA Iran 2021a). Likewise, since the targeted killing of Iran's top general, there are concerns about industrial control system vulnerability within North America (Mitchell 2020).

When turning to space, and threats to the United States, Iran has posed a challenge by its own investment in space-enabled services; this is proven through its implementation of electronic jamming capabilities (Security in Space 2019). These space-enabled services underscore its independent launch capabilities; these can subsequently serve as the basis for a ballistic missile program.

Iran sees value in its investment in jamming capabilities directed toward cyber intensive systems like the global positioning system (Security in Space 2019, 31). While simultaneously engaged in the pursuit of this type of technology, Tehran is also looking to improve its own space-based communications system to add redundant layers to Iran's telecommunications industry (Security in Space 2019). It is also using cyber to weaken U.S. and allied diplomatic influence in today's world (NCS 2018, 2).

Iran's burgeoning cyber investment is a major component of Iran's strategy. Since 2012 and continuing to the present time, Iran has invested extensive resources into its cyber-attack and cyber espionage capabilities (Katzman 2019, 5; CISA Iran 2021a). Accordingly, it is using cyber to conduct espionage and to gain intelligence on U.S. interests inside this country (WTA 2021, 14). Anderson and Sadjadpour (2018, 14) calculated the number of attacks perpetrated by Iran to be in the tens of millions, thus showing a proclivity in using cyber as a significant weapon in Iran's arsenal for the future. Accordingly, Iran sees its adeptness in cyber as a key pillar of the asymmetrical warfare TTPs it uses to thwart the United States and other western allies in the Middle East (Bahgat and Anoushiravan 2017). Both the Islamic Revolutionary Guard Corps and Iranian intelligence services have successfully implemented several cyber-attacks recently (Anderson and Sadjadpour 2018, 5).

It has also invested in other cyber offensive capabilities and has launched cyber operations to influence popular opinion, steal information, and disrupt critical infrastructure (WTA 2019, 5). It is important to note, Iran has conducted critical infrastructure attacks against the United States and other allies and now is capable of creating localized and temporary outcomes which include the likes of network disruption of major corporations for days or weeks which is akin to those the Saudi government experienced in 2016 and 2017 (WTA 2019, 6). Iran's attacks against the United States and its allies have looked to maximize damage (Lewis 2019).

Iran also pursued more specific cyber espionage and attack TTPs in recent years (WTA 2019, 6). It has done so to gather information and to help it leverage derivative advantages from using such asymmetric means (Weaver 2017; Weaver and Johnson 2020). This country demonstrated its ability to target U.S., and international businesses and caused economic harm; it will most likely continue until action is taken to change its behavior (NCS 2018, 1–2). An Iranian organization called the Mabna Institute recently hacked the networks of 144 U.S.-based colleges and universities (Weaver et al. 2020). The United States subsequently identified and indicted Iranians complicit in these incidents (Iran Action Group 2018). This country has been committed to the development of cyber forces to allow it to conduct offensive operations (GAO-19-204SP 2018, i). What is more is that Iran is quite adept at also using social media in targeted campaigns to influence citizens of the United States (and its allies) while it attempts to spread messages aligned with Iranian interests; it will most likely continue to do so as it tries to advance its causes (WTA 2019).

Iran attempted to gain access to the accounts of American representatives recently during nuclear negotiations; this demonstrated the tenacity of this actor's interest in acquiring intelligence to undermine the position of the United States (Anderson and Sadjadpour 2018, 31). Another recent Iranian cyber-attack involved an attempt to acquire data by impersonating a trusted source to increase the likelihood that it could acquire information from members of the United States' Congress about potential sanctions that were being targeted toward Iran (Anderson and Sadjadpour 2018, 31). Moreover, Bennie Thompson, the House Homeland Security Chairman, echoed DHS' concern about the possibility of future Iranian attacks on American critical infrastructure.

When revisiting the YIRTM-C (M) and turning to the evidence in Annex 6.1, Iran showed some differences with regard to what was discovered when looking et al. Qaeda and China. More pointedly, the evidence showed that Iran mostly used the information and economic components of the model almost equally. It has done so to seek advantages over adversaries like the United States (and its allies), to target countries that seek to impose sanctions to adversely impact it economically and Iran also wants to discredit those that it believes want to do it harm. Iran has demonstrated a proclivity in using cyber to gain intelligence on the United States and other western allies.

Additionally, to a relatively equal level (albeit at a level lesser than information and economic means), Iran used cyber to execute influence operations to undermine western diplomatic efforts. Likewise, its investment in weaponized cyber (military) is designed to exploit weaknesses and to determine vulnerabilities in nations' infrastructure to attack or exploit these at a time of its choosing should it feel threatened existentially.

Annex 6.1: Iran

How/Why	D.I.M.E	Source type	Author	Date	Page(s)
Iran has posed a challenge in investing in cyber and space-enabled services proven through jamming capabilities	I, M, E	Government Document	Security in Space	2019	iii

(continued)

(continued)

How/Why	D.I.M.E	Source type	Author	Date	Page(s)
Iran sees advantages through investing in networked jamming capabilities directed at the global positioning system	I, M, E	Government Document	Security in Space	2019	31
It looks to enhance its own space-based communications system to create redundancy to this country's telecommunications industry	I, M, E	Government Document	Security in Space	2019	31
Iran conducted cyber operations to influence popular opinion, steal information, and disrupt critical infrastructure	D, I, E	Government Document	WTA	2018	5
Iran conducted critical infrastructure attacks against the United States; it can cause localized and temporary outcomes which include the likes of the disruption of major corporate networks	I, E	Government Document	WTA	2018	6
Iran is using cyber to acquire information and to leverage advantages through using such asymmetric means	I, M, E	Book Chapter	Weaver	2017	29–54

(continued)

(continued)

How/Why	D.I.M.E	Source type	Author	Date	Page(s)
The former head of the CIA, Gina Haspel spoke recently of the Iranian threat and how this has garnered the attention of the agency in recent years	I, M, E	Speech	Haspel	2019	NP
It uses cyber to weaken U.S. and allied influence in the world today	D	Government Document	NCS	2018	2
Iran has shown its ability to target U.S. and international businesses; it caused economic harm and will most likely continue until action is taken to curb its behavior	E	Government Document	NCS	2018	1–2
Iran is committed to the development of cyber forces to conduct offensive operations	I, M	Government Document	GAO-19-204SP	2018	i
Iran's expansion of cyber investment is a significant component of its grand strategy	I, M, E	Government Document	Anderson and Sadjadpour	2018	5
Iran's recent attacks against the United States and allies have been brazen and look to maximize damage	I, M, E	Government Document	Lewis	2019	NP
Iran recently attempted to gain access to the accounts of American representatives during nuclear negotiations	D, I	Government Document	Anderson and Sadjadpour	2018	31

(continued)

(continued)

How/Why	D.I.M.E	Source type	Author	Date	Page(s)
An Iranian cyber-attack that occurred involved an attempt to acquire data by impersonating a trusted source to better allow it to acquire information from members of the United States' Congress about potential sanctions that were being drafted against Iran	D, I	Government Document	Anderson and Sadjadpour	2018	31
The U.S. Department of Homeland Security has attested to increases in Iran's malicious cyber forays directed at the United States in recent times	I, M	Testimony	Leithauser	2019	NP
Concerns increase about Iran's potential attacks on the North American telecommunications infrastructure	I	Testimony	Leithauser	2020	NP
Since the killing of Iran's top general, there are concerns about industrial control system vulnerability within North America	I, E	Press Release	Mitchell	2020	NP
There is an urgent need to improve cyber awareness and security at all levels in the private sector because of vulnerabilities	I, E	Press Release	Krebs	2020	NP

(continued)

(continued)

How/Why	D.I.M.E	Source type	Author	Date	Page(s)
The House Homeland Security Chairman echoed DHS' concern about future attacks on American critical infrastructure	I, E	Press Release	Weber	2020	NP
Iran has also engaged in disinformation campaigns	I	Government Document	CISA Iran	2021a	NP
It is promoting a pro-Iranian message while it looks to foster anti-U.S. sentiment	I	Government Document	CISA Iran	2021a	NP
Iran was also actively attempting to use cyber to alter the U.S. presidential election in 2021	D	Government Document	CISA Iran	2021b	NP
It has remained committed to disruptive and destructive cyber operations targeting strategic objectives, the financial sector, and energy	E	Government Document	CISA Iran	2021a	NP
It most likely will engage in disinformation campaigns to discredit the United States and to circulate content counter to official U.S. messaging	D	Government Document	WTA	2021	14
It is using cyber to conduct espionage and to gain intelligence on U.S. interests inside this country	I	Government Document	WTA	2021	14

REFERENCES

Anderson, C., and K. Sadjadpour. 2018. *Iran's Cyber Threat: Espionage, Sabotage, and Revenge*. Carnegie Endowment for International Peace.

Bahgat, G., and E. Anoushiravan. 2017. Iran's Defense Strategy: The Navy, Ballistic Missiles and Cyberspace. *Middle East Policy* (September 7): 1.

CIA. 2019. The World Factbook, Iran. https://www.cia.gov/library/publicati ons/resources/the-world-factbook/geos/ir.html. Accessed 17 June 2019.

CISA Iran. 2021a. Increased Geopolitical Tensions and Threats. https://www. cisa.gov/publication/increased-geopolitical-tensions-and-threats. Accessed 28 July 2021.

CISA Iran. 2021b. Statement from CISA Director Krebs on Election Security Announcement. https://www.cisa.gov/news/2020/10/21/statement-cisa-director-krebs-election-security-announcement. Accessed 28 July 2021.

Daigle, T. 2020. Here's How Iran Could Seek Revenge with Cyberattacks on the U.S. *CBC*, January 4. https://www.cbc.ca/news/technology/iran-retali ation-cyberattacks-1.5414882. Accessed 10 March 2020.

GAO-19-204SP. 2018. Report to Congressional Committees National Security Long-Range Emerging Threats Facing the United States as Identified by Federal Agencies. https://www.gao.gov/assets/700/695981.pdf. Accessed 23 July 2019.

INSSG. 2021. Interim National Security Strategic Guidance. https://www.whi tehouse.gov/wp-content/uploads/2021/03/NSC-1v2.pdf. Accessed 17 Aug 2021.

Iran Action Group. 2018. Outlaw Regime: A Chronicle of Iran's Destructive Activities. Retrieved from https://www.state.gov/wp-content/uploads/ 2018/12/Iran-Report.pdf.

Haspel, Gina. 2019. CIA Director Gina Haspel Speaks at Auburn University. https://www.cia.gov/news-information/speeches-testimony/2019-speeches-testimony/dcia-haspel-auburn-university-speech.html. Accessed 17 June 2019.

Katzman, K. 2019. Iran's Foreign and Defense Policies. Congressional Research Service. Retrieved from https://fas.org/sgp/crs/mideast/R44017. pdf. Accessed 26 Sept 2019.

Krebs, Brian. 2019. Krebs Says DHS Warning of Iranian Cyber Threat Builds 'Community' with Private Sector. https://search.proquest.com/docview/225 4413489?pq-origsite=summon. Accessed 28 Feb 2020.

Leithauser, Tom. 2019. DHS Warns of Iranian Cyber Attacks. https://search. proquest.com/docview/2262067198?pq-origsite=summon. Accessed 28 Feb 2020.

Leithauser, Tom. 2020. Reps. Pallone, Doyle Seek Briefings on Iranian Cyber Attacks. https://search.proquest.com/docview/2347763342?pq-ori gsite=summon. Accessed 28 Feb 2020.

Lewis, J. 2019. Iran and Cyber Power. Center for Strategic and International Studies. Retrieved from https://www.csis.org/analysis/iran-and-cyber-power. Accessed 21 Sept 2019.

Mitchell, Charlie. 2020. Industrial Operators on Notice About Enhanced Iranian Cyber Threat. https://search.proquest.com/docview/2334164239?pq-origsite=summon. Accessed 28 Feb 2020.

NCS. 2018. National Cyber Strategy of the United States of America. https://www.whitehouse.gov/wp-content/uploads/2018/09/National-Cyber-Strategy.pdf. Accessed 22 July 2019; and Prosperity in the Digital Age. https://www.securitepublique.gc.ca/cnt/rsrcs/pblctns/ntnl-cbr-scrt-strtg/ntnl-cbr-scrt-strtg-en.pdf. Accessed 23 July 2019.

Security in Space. 2019. Challenges to Security in Space. https://www.dia.mil/Portals/27/Documents/News/Military%20Power%20Publications/Space_Threat_V14_020119_sm.pdf. Accessed 10 June 2019.

Weaver, John M. 2017. Cyber Threats to the National Security of the United States: A Qualitative Assessment (Chapter). In *Focus on Terrorism, vol 15*. Nova Science Publishers.

Weaver, Mark, Josh Brashears, Noah Morton, and Alexander Simons. 2020. What Lurks in the Shadows? Cyber Threats to the United States. In *Contemporary Intelligence Analysis and National Security: A Critical American Perspective*. Hauppauge, NY: Nova Science Publishers.

Weaver, John M., and Benjamin Johnson. 2020. *Cyber Security Challenges Confronting Canada and the United States*. New York, USA: Peter Lang Publishing.

WTA. 2019. Worldwide Threat Assessment of the US Intelligence Community. https://completethreatpreparedness.com/wp-content/uploads/2019/02/2019-ODNI-Worldwide-Threat-Assessment.pdf. Accessed 11 June 2019.

WTA. 2021. Worldwide Threat Assessment of the US Intelligence Community. https://www.dni.gov/files/ODNI/documents/assessments/ATA-2021-Unclassified-Report.pdf. Accessed 3 Aug 2021.

CHAPTER 7

Islamic State (IS)

Abstract Similar to al Qaeda, the Islamic State is a transnational terror organization. As a threat, this chapter explores how this terror organization is using cyber operations to weaken the United States.

Keywords Islamic State · Messaging · YouTube · Emni

Background

The Islamic State has earned the position as being one of the most feared terrorist groups in the contemporary world. IS emerged as an offshoot of the renowned terror group al Qaeda and its leader was Abu Musab al Zarqawi (Hart and Schultz 2019). It formed as a radical but small militant group striving for increased influence and power showing no remorse for the bloodshed it has inflicted, whether or not this included the death of other Muslims. Osama bin Laden, who led al Qaeda, expressed his displeasure over the extremely violent acts that IS conducted on other Muslims; he dictated that his second chief in command at the time, Ayman al-Zawahiri, send Zarqawi a letter instructing him to reduce the collateral damage of attacks, this following the killing of sixty Muslims at a Jordanian wedding using suicide bombers (Hart and Schultz 2019). Zarqawi subsequently disregarded the directive and released his own

J. M. Weaver, *The U.S. Cybersecurity and Intelligence Analysis*, https://doi.org/10.1007/978-3-030-95841-1_7

video claiming himself as the leader of the al Qaeda division in Iraq and at that point was committed to creating an Islamic State (Hart and Schultz 2019).

What followed was a dual leadership structure that included Abu Ayyub al-Masri from Egypt and Abu Umar al Baghdadi from Iraq (Nance 2016). Both leaders were killed in a subsequent American-led airstrike after U.S. counterintelligence sifted through al Qaeda's attempt to conceal this new leadership structure. Once again, the Islamic State struggled to gain momentum. Following a strong American presence in Iraq, al Qaeda began fighting in earnest to maintain its existence (Hart and Schultz 2019).

Once atrophied, the al Qaeda division in Iraq extricated its organization from Iraq and moved across the border to Syria; in the process, it officially named itself ISIL or the Islamic State. Despite the constant leadership and authority changes, the Islamic State pursued an expansionist strategy that became persistently brutal with exceptional violence as it attacked weak points in Iraq (Nance 2016).

Around 2013, IS started attacking other rebel groups to seek greater dominance in the region. In June 2014, it declared IS to be the universal caliphate and its leader, Abu Bakr Baghdadi, the caliph (Hart and Schultz 2019). The Islamic State also asserted that the caliphate holds both religious and political/military power over all Muslims across the planet (Wang et al. 2017). The group then began undertaking extreme measures to instill fear and assert its presence throughout the world with even more brutal and frequent attacks. They seized Mosul and Raqqa and declared the latter to be the capital of the IS; this represented the first major foray into exerting its dominance and influence throughout the Middle Eastern region. During an interview with Matthew Heineman, West and West (2017) wrote on how the once-prosperous Raqqa was now a breeding ground for the terror organization, "...because ISIL had completely blacked out the city from the rest of the world. No information was coming in; no information was going out." The lives of the citizens in Raqqa changed dramatically; they had to decide to stay in the IS-controlled city or move elsewhere (Hart and Schultz 2019).

In late 2018, both Canada's Defence Minister Harjit Sajjan and former U.S. Secretary of Defense James Mattis co-hosted a meeting that included the 13 top nations contributing to the fight against the Islamic State (Sajjan and Mattis 2018). During this session, both attested to the need to create a network to beat a network. While ISIL has lost a great deal of its terrain in Iraq and Syria, there are indications that the organization is regaining strength and growing in tactical sophistication making any analyses of its imminent demise both premature and dangerous (Guerin 2019; Schmitt et al. 2019).

The subsequent analysis will look at the Islamic State and how it is using cyber to improve its use of the D.I.M.E. instruments and by targeting the diplomacy, information, military, and economic resources of the United States through the YIRTM-C (M). More specific evidence drawn from the data is reflected in Annex 7.1. This section looks at the impact on the United States.

ANALYSIS

The Islamic State, as a non-state actor, has been quite adept at using cyber. It uses information technology to (1) market itself as an organization, (2) to communicate with operatives, and (3) spread propaganda. The Islamic State constantly looks for ways to use cyber to conduct mass disruption (GAO-19-204SP 2018, 3). Repeatedly, IS has (and successfully) used internet messaging to promote its cause (WTA 2019, 12). Most likely it will continue to use violent messaging vis-à-vis mainstream media and social media to garner support. Gina Haspel who headed the CIA spoke at a 2019 college graduation about the threat that the Islamic State poses; she stated how this has required greater resourcing of the agency in recent years (Haspel 2019).

People throughout the globe are becoming more susceptible to targeted messaging; many are being persuaded to join the IS movement (Yu and Haque 2016). It is quite adept at using propaganda to spread its message to generate support for its cause (Sajjan and Mattis 2018). This terror organization has expanded its recruitment beyond the Middle East; it has global aspirations, and it desires to grow its ranks. IS continues to make use of social media tactics vis-à-vis videos to persuade westerners of different backgrounds to join the 'brotherhood' (Hart and Schultz 2019). Seemingly, the most impressionable are "the everyday young people who are in social transition, on the margins of society, or amidst in a crisis of identity" (Yu and Haque 2016). Westerners perhaps are uniquely susceptible to recruitment as liberal societies often underscore the notion of individualism, which can also lead to many becoming lonely and they often seek to find some level of community that is attractive through these strategies (Yu and Haque 2016).

The Islamic State possesses firsthand knowledge of the viability in spreading its ideology through cyber and the internet according to Michael Burnett, the U.S. Senior Director of Counterterrorism (2019). According to statements by President Trump in recent years, the Islamic State was even more adept than the U.S. for years in utilizing the internet and cyber as a way to advance the terror organization's position (Trump 2019). Biden is equally concerned with this terror organization (INSSG 2021, 11).

Focusing on specific platforms, the Islamic State has effectively used social media including YouTube and Facebook to target individuals to join its ranks. These "platforms [serve] as magnets that have attracted thousands of views, comments, forums, and posts" (Awan 2017, 139). Since IS began using YouTube, its campaign is worth one billion U.S. dollars centered on using the information to recruit young, impressionable Muslims to join its cause (Hart and Schultz 2019). The Islamic State has gone to great lengths in creating a phone app titled "The Dawn of Glad Tidings" (Hart and Schultz 2019). This app centered on the updates and activities in which they are involved. However, subsequent coalition efforts have detected and then suspended this app. It is incumbent upon the United States to continue its vigilance in monitoring social media platforms like these to counter the Islamic State's information in a timely fashion so that campaigns to reduce the number of people joining IS are more effective at inhibiting its recruitment tactics (Awan 2017).

IS, as a non-state actor, has demonstrated a great deal of resilience. To continue its existence, it has published a cybersecurity survival guide (ISIS 2015). The organization's plan promotes the use of TTPs, and these include disabling mobile phone location services, using encryption, the use of a Russian encrypted messaging application called 'Telegram,' ensuring operational security when taking pictures or making videos, learning how to operate in internet denied environments, as well as providing instruction on how to browse anonymously online.

The primary intelligence component of the Islamic State is a subcomponent referred to as Emni; it has a significant influence on the military component of the YIRTM-C (M). While mostly gravitating around the information and military components of the model, most of Emni's actions support the multitude of factions that comprise the Islamic State and assist it in achieving greater success (Hart and Schultz 2019). Emni emerged to gather information and process it thereby allowing the Islamic State to advance and eventually challenge its enemies. "Among its many tasks, the Emni actively controls and monitors the flow of ISIL's logistical support operations inside Turkey that have been crucial to its operations, including the flow of materials used for explosives (igniters, chemicals, fertilizers, cables, et cetera that have been funded through Turkey to ISIL) and other deliveries critical to them" (Speckhard and Yayla 2016). Likewise, the Islamic State is an organization that realizes the prominence of the internet and all that it provides. More pointedly, IS has been able to use the internet to foster regrouping and in rebuilding its militant capabilities through continuous recruitment (Burnett 2019).

Emni is a significant enabler that will most likely continue to focus on the acquisition of information to gain control of land and weapons so that eventually ISIL will expand its ranks (militarily) (Weaver and Johnson 2020).

Moreover, Emni is essential for the conduct of internal investigations and for improving military recruitment (Hart and Schultz 2019). This sub-organization of IS has the responsibility for actively training its military to return operatives to their homeland to conduct attacks. "It is now understood that Emni trained operatives [that] carried out the 2015 Paris café, stadium, and nightclub attacks as well as recruited the cadres for and built the bombs used in the 2016 Brussels airport and metro attacks" (Speckhard and Yayla 2016). Thus, Emni's adeptness seemingly weakens the United States because of its lack of experience with intelligent terrorist groups (Weaver and Johnson 2020).

Emni has other priorities and responsibilities that focus on weakening its enemies. This helps it garner the strength and resolve of its military units and helps IS become more mission-focused (Weaver and Johnson 2020). Its duties include typical counterintelligence activities like investigating its members to ensure that no insider threats exist, thus helping to ensure those opposed to its operations are not likely to infiltrate its units. "On the technical side, its computers are also monitored in a low-tech manner for internet history and with free downloadable apps that allow monitors to know if banned sites are being accessed" (Speckhard and Yayla 2016). Emni also plans at a strategic level, employs its military TTPs using cyber, and its top leadership identifies who among western IS personnel are to be sent back (Hart and Schultz 2019). IS also looks to specify targets and organize the logistics support for operatives, including paying smugglers to get them to their target locations and, according to European intelligence documents, in at least one case, using cyber for transmitting Western Union transfers (Speckhard and Yayla 2016). This

TTP helps ensure attacks are being executed by the most trustworthy and loyal militant followers that will not raise suspicion in western countries (Hart and Schultz 2019). The complexity and sophistication of Emni's priorities that it is establishing is a significant threat to the United States because even though extremists are often perceived by most westerners to be relatively unorganized compared to state-based actors, IS strategically plans and implements significant, coordinated, and violent attacks even though it has not been as effective in recent years (WTA 2021, 23).

In October 2019, the U.S. killed IS' leader Abu Bakr al-Baghdadi (Weaver and Johnson 2020). IS can still inspire attacks, though weakened, and use information technology to keep its ideology alive (Cronk 2019). Though hurt, it is still a challenging force.

Moreover, Klausen (2017) makes the argument that IS makes good use of social media to implement war tactics throughout the globe. Like YouTube and Facebook, the terror organization has been able to use Twitter to create the illusion that the organization is much larger and more capable than it is (Awan 2017). Awan's (2017) research showed that the United States was in the top 10 countries where IS has been successful at garnering support for its cause, mostly through the use of cyber. This demonstrates how difficult counter-radicalization efforts are for the United States (Weaver and Johnson 2020).

A major instrument that IS relies on to support its operations and development of TTPs is its finances. The Islamic State was quite skilled in generating income even before becoming one of the major terrorist groups in the world. The former U.S. Treasury Under Secretary for Terrorism and Financial Intelligence, David Cohen, stated that ISIL was "The best-funded terrorist organization we've confronted" (Stergio 2016).

Estimates point to IS generating significant revenue totaling between $70 and $200 million a year from its conduct of illicit activities like drug trafficking and extortion (Hart and Schultz 2019). The Islamic State acquired significant shares in the Iraqi oil industry, controlled many fueling stations in northern Iraq, and was actively taking money from businesses in recent years. Speculation surfaced that most of the revenue generated by IS came from five main sources: illicit proceeds from occupation territories, kidnapping for ransom, material support, donations from individuals sympathetic to its cause including by or through non-profit organizations, and lastly, fundraising using modern communication networks (like social media) (Stergio 2016).

Conversely, coalition partners have implemented steps aimed to stop this process and have been successful (Global Coalition 2018). More pointedly, the Global Coalition has launched efforts directed at 30 banks and financial centers that IS was using (Hart and Schultz 2019).

The evidence in Annex 7.1 shows what the Islamic State has been doing to remain effective in terms of the use of the instruments reflected in the YIRTM-C (M). This non-state actor has demonstrated through its reliance on and support of Emni that it is committed to using cyber to draw foot soldiers to its cause, conduct influence operations, and grow its ranks by creating glossy propaganda videos of successful exploits; accordingly, it has mostly used cyber to show its ability to conduct information and military operations to either directly or indirectly harm the United States.

ISIL also has shown a proclivity to exploit the economic component through cyber to gain access to significant sums of money to conduct operations. To the least extent (albeit more effective than al Qaeda), IS has been able to use cyber TTPs to reach out and strengthen alliances with like-minded terror organizations like Boko Haram.

ANNEX 7.1: ISLAMIC STATE

How/Why	D.I.M.E	Source type	Author	Date	Page(s)
The Islamic State has frequently used internet messaging as a way to promote its cause	I, M	Government Document	WTA	2018	12
Gina Haspel, who headed the CIA, spoke at a college graduation about the threat that the Islamic State poses	I, M, E	Speech	Haspel	2019	NP
Emni strategically plans its military tactics using cyber	I, M, E	Book Chapter	Hart and Schultz	2019	263

(continued)

(continued)

How/Why	D.I.M.E	Source type	Author	Date	Page(s)
IS chooses targets and organizes the logistics support for operatives, using cyber conveyances like sending Western Union transfers	I, M, E	Journal	Speckhard and Yayla	2016	2–16
While mostly centered on the information and military components, most of Emni's actions support each faction of the Islamic State	I, M, E	Book Chapter	Hart and Schultz	2019	276
The world is more susceptible to targeted messaging and people are being persuaded to join the IS movement; this terror organization does not limit its recruiting to the Middle East; it has global aspirations, and it has expanded its recruitment to grow its ranks	I	Journal	Yu and Haque	2016	1–6
The Islamic State has successfully used social media platforms in targeting individuals to join its cause; these "platforms [are] as magnets that have attracted thousands of views, comments, forums, and post"	I, M	Journal	Awam	2016	138–149

(continued)

(continued)

How/Why	D.I.M.E	Source type	Author	Date	Page(s)
Research showed that the United States was in the top 10 countries where IS has successfully gained support for its cause	I, M	Journal	Awam	2016	138–149
The organization is using propaganda to promote its message in order to garner support to its cause	I	Press Release	Sajjan and Mattis	2018	NP
IS has the ability to continue to inspire attacks and use information technology to keep its ideology alive	I, M	Press Release	Cronk	2019	NP
The Islamic State has firsthand knowledge of the viability in how it is able to spread its ideology through the internet and cyber	I	Press Release	Burnett	2019	NP
The Islamic State realizes the power that the internet provides; it has been able to use the internet to regroup and rebuild its militant capabilities	M	Press Release	Burnett	2019	NP
President Trump stated that the Islamic State was even more adept than the United States for years in using the internet and cyber as a way to advance their position	I	Speech	Trump	2019	NP

(continued)

(continued)

How/Why	D.I.M.E	Source type	Author	Date	Page(s)
The Islamic State constantly looks for ways to use cyber to conduct mass disruption	I, M, E	Government Report	GAO-19-204SP	2018	3
A recent CSIS report showed how the Islamic State actively pursues the use of social media to promote its terror recruitment and terror messaging	D	Government Report	CSIS Report	2018	20
The Islamic State has helped avoid detection by using alternate internet platforms like the dark web in order to share links	D	Government Report	CSIS Report	2018	20
As the Islamic State looks to remain relevant and expand its influence, the organization will most likely continue to use encryption	D	Government Report	CSIS Report	2018	20

References

Awan, Imran. (2017). Cyber-Extremism: Isis and the Power of Social Media. *Society* 54 (2): 138–149.

Burnett, Michael. 2019. Winning the War against Terror Messaging. https://www.whitehouse.gov/articles/winning-war-terrorist-messaging/. Accessed 22 July 2019.

Cronk, Terri Moon. 2019. ISIS Continues to Pose a Significant Threat, OIR Official Says. https://dod.defense.gov/News/Article/Article/1847711/isis-continues-to-pose-significant-threat-oir-official-says. Accessed 21 July 2019.

CSIS Report. 2018. 2018 CSIS Public Report. https://www.canada.ca/content/dam/csis-scrs/documents/publications/2018-PUBLIC_REPORT_ENGLISH_Digital.pdf. Accessed 21 Aug 2019.

GAO-19–204SP. 2018. Report to Congressional Committees National Security Long-Range Emerging Threats Facing the United States as Identified by Federal Agencies. https://www.gao.gov/assets/700/695981.pdf. Accessed 23 July 2019.

Global Coalition. 2018. Mission Tackling Daeshs Financing and Funding. Accessed 26 Sept 2018. http://theglobalcoalition.org/en/mission-en/#tackling-daeshs-financing-and-funding.

Guerin, O. 2019. Isis in Iraq: Militants 'Getting Stronger Again'. *BBC*, December 23. https://www.bbc.com/news/world-middle-east-50850325. Accessed 10 Mar 10 2020.

Hart, Alexis, and Brielle Schultz. 2019. Monsters of the Middle East: ISIL's Perpetual Pursuit of Power in the Middle East (A Qualitative Assessment) (Chapter). In *Global Intelligence Priorities (from the Perspective of the United States)*. Nova Science Publishers.

Haspel, Gina. 2019. CIA Director Gina Haspel Speaks at Auburn University. https://www.cia.gov/news-information/speeches-testimony/2019-speeches-testimony/dcia-haspel-auburn-university-speech.html. Accessed 17 June 2019.

INSSG. 2021. Interim National Security Strategic Guidance. https://www.whitehouse.gov/wp-content/uploads/2021/03/NSC-1v2.pdf. Accessed 17 Aug 2021.

ISIS. 2015. Several Cybersecurity to Protect Your Account in the Social. https://www.wired.com/wp-content/uploads/2015/11/ISIS-OPSEC-Guide.pdf. Accessed 7 Mar 2016.

Klausen, J. 2017. Tweeting the Jihad: Social Media Networks of Western Foreign Fighters in Syria and Iraq. *Studies in Conflict & Terrorism* 38: 1–22.

Nance, Malcolm. 2016. *Defeating ISIS: Who They Are, How They Fight, What They Believe*. New York, NY: Skyhorse Publishing, ProQuest Ebook Central. https://ebookcentral.proquest.com/lib/ycp/detail.action?docID=5304584.

Sajjan, Harjit, and James Mattis. 2018. Canada-U.S. Joint Statement. https://dod.defense.gov/News/News-Releases/News-Release-View/Article/1706457/canadaus-joint-statement/. Accessed 21 July 2019.

Schmitt, E., A.J. Rubin, and T. Gibbons-Neff. 2019. ISIS is Regaining Strength in Iraq and Syria. *The New York Times*, August 19. https://www.nytimes.com/2019/08/19/us/politics/isis-iraq-syria.html. Accessed 10 Mar 2020.

Speckhard, Aanne, and Ahmet S. Yayla. 2016. The ISIS Emni: The Inner Workings and Origins of ISIS's Intelligence Apparatus. The International Center for the Study of Violent Extremism, December 3. *Perspectives on Terrorism* 11 (1): 2–16.

Stergio, Dimitrios. 2016. ISIS Political Economy: Financing a Terror State. *Journal of Money Laundering Control* 19 (2): 189–207.

Trump, Donald. 2019. Remarks by President Trump to the Ministers of the Global Coalition to Defeat ISIS. https://www.whitehouse.gov/briefings-statements/remarks-president-trump-ministers-global-coalition-defeat-isis/. Accessed 22 July 2019.

Wang, Xingang, Zhang Wentao, and Yang Yulong. 2017. Ideology, Global Strategy, and Development of the Islamic State and Its Influence on China's "One Belt, One Road" Initiative. *Journal of Global South Studies* 34 (2): 139–155.

Weaver, John M., and Benjamin Johnson. 2020. *Cyber Security Challenges Confronting Canada and the United States.* New York, NY: Peter Lang Publishing.

West, Dennis, and Joan M. West. 2017. Seeking the Truth about Raqqa: An Interview with Matthew Heineman. https://web.b.ebscohost.com/ehost/detail/detail?vid=0&sid=42e832e4%E2%80%933954%E2%80%934ebd-a02c-b90f8dbc34f4%40sessionmgr102&bdata=JnNpdGU9ZWhvc3QtbGl2ZSZz Y29wZT1zaXRl#AN=124993569&db=a9h. Accessed 3 Oct 2018.

WTA. 2019. Worldwide Threat Assessment of the US Intelligence Community. https://completethreatpreparedness.com/wp-content/uploads/2019/02/2019-ODNI-Worldwide-Threat-Assessment.pdf. Accessed 11 June 2019.

WTA. 2021. Worldwide Threat Assessment of the US Intelligence Community. https://www.dni.gov/files/ODNI/documents/assessments/ATA-2021-Unclassified-Report.pdf. Accessed 3 Aug 2021.

Yu, Seong Hun, and Omar Sultan Haque. 2016. Vulnerabilities among Young Westerners Joining ISIS. *Brown University Child and Adolescent Behavior Letter* 32 (2): 1–6.

CHAPTER 8

Democratic People's Republic of Korea (DPRK/North Korea)

Abstract When turning to Asia, one rogue actor emerges that is a threat to the United States. This chapter explores how North Korea is using cyber operations to weaken the United States.

Keyword North Korea · Kim Jong-un · APT · WannaCry

Background

Following World War II, the Korean peninsula was separated with the northern portion coming under communist control (Soviet-sponsored) (CIA 2019). After the DPRK failed to retake South Korea in the Korean War (1950–1953), North Korea, under its founder President Kim Il Sung, subscribed to a policy of diplomatic and economic 'self-reliance' as a hedge against external influence (CIA 2019). Pyongyang is its capital, and the country celebrated its independence from Japanese imperial rule as of August 15, 1945. The legal system of North Korea is one of civil law based upon the Prussian system but is influenced by Japanese traditions as well as the legal theory of the communists (CIA 2019).

J. M. Weaver, *The U.S. Cybersecurity and Intelligence Analysis*, https://doi.org/10.1007/978-3-030-95841-1_8

Though it is a single-party dictatorship, its governmental structure is made up of the executive, legislative, and judicial branches (CIA 2019). Under its executive branch, the Supreme People's Assembly President Choe Ryong Hae is the chief of state (since April of 2019). That stated, the head of government is the real person in charge, which is Chairman Kim Jong-un; he has ruled since December of 2011 (CIA 2019).

When turning to the legislative branch, North Korea's is a unicameral Supreme People's Assembly (CIA 2019). There are 687 seats where its members are directly elected by a majority in up to two rounds (if required) for five-year terms (CIA 2019).

Lastly, the judicial branch consists of a Supreme Court or Central Court which has one judge and two 'People's Assessors' or during some instances three judges (CIA 2019). These judges are elected for five-year terms by the Supreme People's Assembly (CIA 2019).

North Korea remains one of the world's most closed and centrally directed economies; it is beleaguered by chronic economic problems (CIA 2019). The DPRK's industrial capabilities are almost beyond repair as a direct result of significant years of underinvestment, shortages of repair parts, and poor infrastructure maintenance (Weaver and Johnson 2020). The country's investment in a major military establishment and the development of North Korea's ballistic missile/nuclear program severely depletes its resources that would be needed for investment and civilian consumption (CIA 2019). The power and industrial outputs of North Korea have been stymied for decades and remain at just a fraction of the levels seen before the 1990s (Weaver and Johnson 2020). Crop failures resulting from weather-related conditions have aggravated its chronic food shortages that have been experienced by its general population that stems from insufficient fertilization, a lack of arable land, bad collective farming practices, poor soil quality, and persistent shortages of farming equipment (like tractors), and the lack of availability of fuel (CIA 2019).

Beginning in the mid-1990s and continuing through the mid-2000s saw severe famine and widespread starvation in the DPRK (CIA 2019). The international community subsequently provided food aid through 2009. Since then, food assistance has dropped off precipitously (CIA 2019). Though internal rice and corn farming have improved in recent times, domestic production has not led to self-sufficiency (CIA 2019).

The vast majority of North Koreans continue to suffer from prolonged malnutrition and poor living conditions (CIA 2019). Since the turn of the millennium, the government has allowed for semi-private markets to emerge enabling the selling of a varied range of goods, thus allowing North Koreans to make up in part for its diminished distribution system of public rations (CIA 2019). North Korea has also implemented changes in the way communal farms are managed to better improve agricultural output (CIA 2019).

As of December 2009, this country carried out a redenomination of its currency that ended up capping the amount of North Korean won that could be exchanged for the new notes; it also limited the exchange to a one-week window (CIA 2019). A crackdown occurred simultaneously, and the use of foreign currency yielded severe shortages of goods and services and this led to inflation, thus pushing Pyongyang to ease the restrictions by February 2010 (CIA 2019). In direct response to the sinking of a South Korean warship, the Cheonan, and the shelling of Yeonpyeong Island in 2010, South Korea's government suspended most of its aid, ceased bilateral cooperation activities, and stopped trade (CIA 2019). In February 2016, South Korea halted what remained of its bilateral economic activity by shutting down the Kaesong Industrial Complex; this was in direct response to the DPRK's fourth nuclear test (CIA 2019). It was this test and another that occurred in September of that year that resulted in two United Nations Security Council resolutions that focused on North Korea's foreign currency earnings, with a particular targeting of its mining industry centered on coal and other mineral exports (Weaver and Johnson 2020). North Korea subsequently continued missile and nuclear tests through 2017; this led to a tightening of United Nations sanctions that saw full sectoral bans on DPRK exports and drastically limited key imports. China has been North Korea's primary trading partner and lifeline over the last decade (CIA 2019).

North Korea possesses one of the largest militaries in the world through its readiness level is called into question. It consists of the Korean People's Army (KPA) which is made up of an Air Force, Ground Forces, a Navy, an Air Force, and its civil security forces (CIA 2019). Moreover, the DPRK has invested a lot in weapons of mass destruction or WMDs, special operations capabilities, cyber, and weapons for electromagnetic and electronic warfare (Tasic 2019; Weaver et al. 2020).

North Korea has seen the flexing by the international community on it and in part, due to its economic failures, has mostly focused on improving its cyber capabilities (Ford 2017). The subsequent analysis looks at North

Korea and how it is making use of cyber to enhance its position through the YIRTM-C (M) and by targeting the diplomacy, information, military, and economic resources of the United States. Specific evidence from the data is reflected in Annex 8.1.

ANALYSIS

At present, Biden is concerned with North Korea's technical prowess and its ability to meddle in the affairs of the United States (INSSG 2021, 8). Accordingly, the DPRK is a near failed state; it lacks many distinct advantages. Countries that are disadvantaged often turn to cyber advanced persistent threats (APTs) to take advantage of asymmetric TTPs (Springer 2017). That stated, the Worldwide Threat Assessment (WTA 2019) has addressed the idea that North Korea is a significant cyber threat to the United States, and by extension its allies, and describes the ability of this country to implement cyber espionage, and disruptive cyber-attacks that enable it to steal millions from financial institutions to generate hard currency. Likewise, it actively seeks to improve its cyber operations and strategy (Jun et al. 2016). It is utilizing cyber to help counter regional nations (namely Japan and South Korea) and globally it looks to identify weaknesses in infrastructure (Weaver and Johnson 2020). It believes that by doing so will enable the country to more aptly counter the United States asymmetrically (Bennett et al. 1999).

In the *Challenges to Security in Space* document, and as was seen with Iran, North Korea presents a clear threat to North America, by its investment in a space-based program designed to enhance jamming capabilities (Security in Space 2019). It does not rely on other nations and has independent launch capabilities that can be seen as a point of departure for a ballistic missile program. North Korea realizes the advantage that space offers and will look to deny potential adversaries the ability to use space freely (Security in Space 2019). The United States has a proclivity in using and relying on space-based systems for NORAD to effectively operate (Weaver and Johnson 2020).

When shifting gears to cyber specifically, and like Iran, North Korea is increasing its use of cyber TTPs to steal information, influence people, and disrupt critical infrastructure (WTA 2019, 5). It is committed to using cyber to hack into the financial sector (CISA North Korea 2020).

More pointedly, it has been involved in a program to steal cryptocurrency (CISA North Korea 2021; WTA 2021, 16). It has used cyber specifically as a way to harm U.S. and partner influence on the global stage (NCS 2018, 2). More specifically, North Korea targets financial institutions to help it generate revenue (hard cash), which has included the attempt to illegally gain access to billions of dollars from institutions across the globe with one specific example pointing to a successful heist of approximately $81 billion from the New York Federal Reserve (WTA 2019, 6). It will most likely continue to target and exploit weaknesses to cause disruption with businesses inside the United States (WTA 2021, 15). Likewise, the information infrastructure and cyberspace networks of the U.S. have been vulnerable and thus becoming low-hanging fruit by state actors including North Korea (Pomeroy 2019). At present, cyber activities by adversaries of the United States have eroded the U.S. military advantage and negatively impacted the United States' economic prosperity (Pomeroy 2019).

North Korea has been actively engaged in a significant number of hostile acts directly or indirectly oriented at the United States since the 1950s (McDowell et al. 2019). Relatively new to this world of potential threats, however, is the DPRK's relentless cyber realm that can be directed against any nation (McDowell et al. 2019). What is more is that North Korea understands that it must buttress its military with cyber to hedge in its favor (Kwon 2010, 163). North Korea could turn to cyber to exert influence that falls below the threshold of what could trigger a conventional military engagement (Cha 2002).

North Korea's weaponization of cyber and exploitation of the internet has been advantageous. According to the Worldwide Threat Assessment (WTA 2019, 6), "North Korea poses a significant cyber threat to financial institutions, remains a cyber-espionage threat, and retains the ability to conduct disruptive cyber-attacks." The cyber activity of North Korea has hurt the economic interests of the United States as well as private sector businesses in the international community (NCS 2018, 1).

North Korea recently used a malware virus known as WannaCry (a ransomware attack) to attack the United States and other nations (McDowell et al. 2019). This one infected hundreds of thousands of computers (Jasper 2016; McMillan et al. 2017; Clarke and Youngstein 2017). The virus rendered many computers useless; the attacks focused on critical infrastructure gravitating around schools, businesses, hospitals, and homes in as many as 150 countries (McDowell et al. 2019). As one of the primary victims, the United States felt the impact of this malware's capability on electric companies, power grids, hospitals, banks, schools, and personal homes (McDowell et al. 2019). Sony Entertainment, particularly, was one of the main victims of this attack. Many of Sony's workstations were targeted and damaged beyond repair; it resulted in them having to be replaced (Congressional Research Service 2017).

Another signature (of North Korea's cyber-attacks) was that most of its earlier ones were more disruptive than destructive in nature (Congressional Research Service 2017). However, North Korea's capabilities of its cyber TTPs have increased such that the state can use cyber tools that are now more destructive rather than disruptive, thereby prodding other nations to take immediate actions. It is imperative that government officials the world over be vigilant in building redundancy and resilience into their networks to better respond to both disruptive and destructive future cyber-attacks (McDowell et al. 2019).

When considering the evidence in Annex 8.1 and the analysis of data through the lens of the YIRTM-C (M) point to the DPRK leveraging the asymmetric advantages of cyber to enhance its economic position while also attempting to weaken the economies of industrialized countries like the United States. This is in part a corollary response to severe economic hardship experienced by and the imposition of crippling sanctions placed on North Korea in recent times. In particular, Kim Jong-un's regime has demonstrated a penchant for success in targeting financial institutions of many western nations in recent years.

Similarly, North Korea used the information and military components of the YIRTM-C (M) through cyber to hedge its position in relation to other more industrialized countries. Likewise, it sees the value in exploiting information through cyber as a way to enhance its acquisition of information and technology to improve its space-based systems and CNE and CNA activities (Weaver and Johnson 2020).

North Korea is using cyber to enhance its image diplomatically on the world stage. That stated, the DPRK has not been as effective using this instrument in conjunction with cyber as it has with the information, military, and economic components of the YIRTM-C (M).

ANNEX 8.1: NORTH KOREA

How/Why	D.I.M.E	Source type	Author	Date	Page(s)
North Korea is a threat to the United States, by its investment in a space-based program to enhance jamming capabilities	I, M, E	Government Document	Security in Space	2019	iii
North Korea sees advantages that networked space-based operations offer and will look to deny potential adversaries the ability to use space	I, M	Government Document	Security in Space	2019	32
North Korea, like Iran, is using cyber TTPs to influence citizens, steal information, and disrupt critical infrastructure	D, I, E	Government Document	WTA	2018	5
Another area that North Korea is using cyber for is the targeting of the financial sector to help it generate revenue and hard cash	I, E	Government Document	WTA	2018	6
The U.S. information infrastructure and cyberspace networks are more vulnerable to North Korean attacks	I, M, E	Book Chapter	Pomeroy	2019	319
Cyber activities by the U.S. adversaries have eroded the U.S. military advantages; these have adversely impacted the U.S. economic prosperity	M, E	Book Chapter	Pomeroy	2019	319
North Korea is relentless in the cyber realm	I, M	Book Chapter	McDowell, Walker, and Meyers	2019	290

(continued)

(continued)

How/Why	D.I.M.E	Source type	Author	Date	Page(s)
Recently, North Korea has used WannaCry for attacking the United State and other nations	I, E	Book Chapter	McDowell, Walker, and Meyers	2019	292
The virus rendered many of computers useless; these included the likes of schools, businesses, and more in as many as 150 countries	I, E	Book Chapter	McDowell, Walker, and Meyers	2019	292
The United States felt the impact of this malware's capability on power grids, banks, electric companies, schools, hospitals, and personal homes	I, E	Book Chapter	McDowell, Walker, and Meyers	2019	292
Government officials need to know how countries will be able to respond to both disruptive and destructive cyber-attacks	I, M, E	Book Chapter	McDowell, Walker, and Meyers	2019	290–292
It has used cyber to harm U.S. and allied influence on the word stage	D	Government Document	NCS	2018	2
North Korea's cyber activity hurts the economic interests of the United States as well as other private sector businesses in the international community	E	Government Document	NCS	2018	1
The WTA addresses the notion that North Korea is a significant cyber threat to the United States and its allies	I	Government Document	WTA	2019	6

(continued)

(continued)

How/Why	D.I.M.E	Source type	Author	Date	Page(s)
North Korea has recently turned to the internet; moreover, according to the WTA, "North Korea poses a significant cyber threat to financial institutions, remains a cyber espionage threat, and retains the ability to conduct disruptive cyber-attacks"	I, E	Government Document	WTA	2019	6
It is committed to using cyber to hack into the financial sector	E	Government Document	CISA North Korea	2021	NP
More pointedly, it has been involved in a program to steal cryptocurrency	E	Government Document	WTA	2021	14
It will most likely continue to target and exploit weaknesses to cause disruption with businesses inside the United States	E	Government Document	WTA	2021	15

REFERENCES

Bennett, Bruce W., Christopher P. Twomey, and Gregory F. Treverton. 1999. *What Are Asymmetric Strategies?* Santa Monica: Rand Publishing.

Cha, Victor D. 2002. Hawk Engagement and Preventive Defense on the Korean Peninsula. *International Security* 27 (1): 40–78.

CIA. 2019. The World Factbook, North Korea. https://www.cia.gov/library/publications/resources/the-world-factbook/geos/kn.html. Accessed 17 June 2019.

CISA North Korea. 2020. The United States Issues an Advisory on North Korean Cyber Threats. https://www.cisa.gov/news/2020/04/15/united-states-issues-advisory-north-korean-cyber-threats. Accessed 28 July 2021.

CISA North Korea. 2021. CISA, FBI, and Treasury Expose Latest Tool in North Korea's Cryptocurrency Theft Scheme—APPLEJEUS. https://www.cisa.gov/news/2021/02/17/cisa-fbi-and-treasury-expose-latest-tool-north-koreas-cryptocurrency-theft-scheme. Accessed 28 July 2021.

Clarke, Rachel, and Taryn Youngstein. 2017. Cyberattack on Britain's National Health Service—A Wake-Up Call for Modern Medicine. *The New England Journal of Medicine* 377: 409–411.

Congressional Research Service. 2017, August 7. *North Korean Cyber Capabilities: In Brief.* Congressional Research Service. https://fas.org/sgp/crs/row/R44912.pdf. Accessed 23 Sept 2018.

Ford, G. 2017. The Pyongyang Paradox. *Soundings* 67: 135–146.

INSSG. 2021. *Interim National Security Strategic Guidance.* https://www.whitehouse.gov/wp-content/uploads/2021/03/NSC-1v2.pdf. Accessed 17 Aug 2021.

Jasper, Scott E. 2016. North Korea's Cyberspace Aggression. *International Journal of Intelligence and Counterintelligence* 32 (1): 194–198.

Jun, Jenny, Scott LaFoy, and Ethan Sohn. 2016. *North Korea's Cyber Operations: Strategy and Response.* Washington, DC: Center for Strategic & International Studies.

Kwon, Yang-ju. 2010. *The Comprehension of North Korean Military.* Seoul: Korea Institute of Defense Analysis.

McDowell, Nathan, Ethan Walker, and Matthew Meyers. 2019. Prominent Cybersecurity Issues for the United States: A Qualitative Assessment (Chapter). In *Global Intelligence Priorities (from the Perspective of the United States)*, ed. Jennifer Y. Pomeroy and John M. Weaver. New York: Nova Science Publishers.

McMillan, Robert, Jenny Gross, and Denise Roland. 2017. Major Cyberattack Sweeps Globe Hitting FEDEX, U.K. Hospitals, Spanish Companies. *The Wall Street Journal*, May 12.

NCS. 2018. National Cyber Strategy of the United States of America. https://www.whitehouse.gov/wp-content/uploads/2018/09/National-Cyber-Strategy.pdf. Accessed 22 July 2019 and Prosperity in the Digital Age. https://www.securitepublique.gc.ca/cnt/rsrcs/pblctns/ntnl-cbr-scrt-strtg/ntnl-cbr-scrt-strtg-en.pdf. Accessed 23 July 2019.

Pomeroy, Jennifer Yongmei. 2019. Challenges of the U.S. National Security and Moving Forward (Chapter). In *Global Intelligence Priorities (from the Perspective of the United States)*, ed. Jennifer Y. Pomeroy and John M. Weaver. New York: Nova Science Publishers.

Security in Space. 2019. *Challenges to Security in Space.* https://www.dia.mil/Portals/27/Documents/News/Military%20Power%20Publications/Space_Threat_V14_020119_sm.pdf. Accessed 10 June 2019.

Springer, Paul J. 2017. *Encyclopedia of Cyber Warfare*. Santa Barbara: ABC-CLIO, LLC.

Tasic, M. 2019. Exploring North Korea's Asymmetric Military Strategy. *Naval War College Review* 72: 59–77.

Weaver, John M., and Benjamin Johnson. 2020. *Cyber Security Challenges Confronting Canada and the United States*. New York, USA: Peter Lang Publishing.

Weaver, Mark, Josh Brashears, Noah Morton, and Alexander Simons. 2020. What Lurks in the Shadows? Cyber Threats to the United States. In *Contemporary Intelligence Analysis and National Security: A Critical American Perspective*, ed. John M. Weaver and Jennifer Y. Pomeroy. New York: Nova Science Publishers.

WTA. 2019. *Worldwide Threat Assessment of the US Intelligence Community*. https://completethreatpreparedness.com/wp-content/uploads/2019/02/2019-ODNI-Worldwide-Threat-Assessment.pdf. Accessed 11 June 2019.

WTA. 2021. *Worldwide Threat Assessment of the US Intelligence Community*. https://www.dni.gov/files/ODNI/documents/assessments/ATA-2021-Unclassified-Report.pdf. Accessed 3 Aug 2021.

CHAPTER 9

Russian Federation (Russia)

Abstract Russia is currently seen as the second most pressing threat to the national security of the United States. Accordingly, this chapter goes into great detail on how it is using cyber operations to weaken the United States.

Keywords Russia · Putin · GRU

BACKGROUND

Like one saw with China, Russia is among the more advanced state actors using cyber to exploit vulnerabilities in the United States to its advantage. The capital of Russia is Moscow, and the country is a semi-presidential federation (CIA 2019). Its Independence Day fell on December 25, 1991 (from the former Soviet Union); Russia's legal system is one underpinned by a civil law system with judicial reviews of its legislative acts (CIA 2019).

The head of Russia's executive branch is the chief of state, President Vladimir Putin, and the head of government is Mikhail Vladimirovich Mishustin. Putin re-assumed the presidency in May of 2012 (CIA 2019).

The legislative branch is a Federal Assembly (bicameral). The Soviet Federatsii, or Federation Council, consists of 170 seats, and the State Duma or Gosudarstvennaya Duma has 450 seats (CIA 2019).

J. M. Weaver, *The U.S. Cybersecurity and Intelligence Analysis*, https://doi.org/10.1007/978-3-030-95841-1_9

Under its judicial branch, the Russian Federation's highest court is the Supreme Court and has 170 members. The members of the highest three courts in Russia are nominated by the president; they are subsequently appointed by the Federation Council for life (CIA 2019). Russia has several subordinate courts, and these are as follows: The Higher Arbitration Court, regional (kray) and provincial (oblast) courts, the Moscow and Saint Petersburg city courts, and finally the autonomous province and district courts (CIA 2019).

Russia has seen major changes since the collapse of the Soviet Union thus serving as an impetus moving it from a centrally planned economy toward one that is more market-based (CIA 2019). Both economic growth and political reform have stalled recently and as a result, it remains a predominantly statist economy where a high concentration of wealth is in the hands of officials (CIA 2019). Economic reforms throughout the mid-to-late 1990s privatized most industries, which was largely taken over by a small group of oligarchic elites; there have been notable exceptions in the energy, banking, transportation, and defense sectors, which remain under the purview of the state (CIA 2019). Russia's protection of property rights remains weak, and the state often interferes in the free operation of its private sector (CIA 2019).

Conversely, Russia is one of the global leading producers of oil and natural gas and is also a major exporter of metals (primarily aluminum and steel) (CIA 2019). It is heavily dependent on the movement of commodity prices on the world markets as reliance on these types of exports makes it vulnerable to expansion and contraction cycles that often follow with volatile swings in global prices (CIA 2019). The economy has seen diminished growth rates since 2008 primarily due to the exhaustion of Russia's commodity-based growth model (CIA 2019).

Through the intersection of falling oil prices, structural limitations, and international sanctions these pushed Russia into a deep recession in 2015, with its GDP sinking by almost 2.8%. This downturn continued well into 2016, with GDP contracting another 0.2%, but was reversed back in 2017 as global demand for Russian resources increased (CIA 2019). This government's support for import substitution has seen increases in recent years to diversify the economy away from extractive industries (CIA 2019).

Russia's military is quite formidable (Weaver and Johnson 2020). It consists of Ground Troops (Sukhoputnyye Voyskia, SV), a Navy

(Voyenno-Morskoy Flot, VMF), an Aerospace Force (Vozdushno-Kosmicheskiye Sily, VKS), Airborne Forces (Vozdushno-Desantnyye Voyska, VDV) in addition to Missile Troops of Strategic Purpose (Raketnyye Voyska Strategicheskogo Naznacheniya, RVSN) which are commonly referred to as Strategic Rocket Forces (that are considered independent 'combat arms,' and are not subordinate to any of the three branches) (CIA 2019).

The subsequent analysis turns to Russia and how it is making use of cyber thereby improving its position through the D.I.M.E. instruments and while also targeting the diplomacy, information, military, and economic resources of the United States through the YIRTM-C (M). More evidence of what is occurring is reflected in Annex 9.1.

ANALYSIS

Like China, Biden sees Russia as a formidable rival (INSSG 2021, 6). Russia's annexation of Crimea and its involvement in the Ukrainian crisis (in addition to the full invasion of Ukraine in late February, 2022), along with its reported interference in the 2016 and 2020 U.S. Federal elections, have created questions on both sides of the Atlantic concerning western states and defense institutions (and their ability) to counter hybrid threats to project stability. Along this line, the vast majority of the literature on policy and military affairs from recent years discuss Russia's use of hybrid tactics that have paired conventional military capabilities with unconventional strategies (e.g., Hoffman 2009; Reichborn-Kjennerud and Cullen 2016). These strategies often include non-uniformed military personnel use along with disinformation campaigns instigated through cyber and media channels in a coordinated fashion. This combination of these tactics and resources was effective in limiting Russia's exposure to criticism.

Hybridization is not necessarily 'new' in this sense, since Russia and the former Soviet Union often are credited with the normalization of disinformation campaigns directed at their enemies. The use of disinformation, traditionally, was used to add to the 'fog of war' in kinetic operations. However, disinformation's use evolved to be employed in broader social contexts. A disinformation campaign example includes Soviet efforts to implicate the United States in the emergence of the AIDS pandemic during the 1980s (Boghardt 2009). Disinformation tactics, more recently, enabled by technology (such as digital alteration tools and 'bots') are used increasingly to influence political outcomes, including altering public opinion and election outcomes (Chesney and Citron 2019). Overall, indicators show that Russia (along with other state actors) is using cyber capabilities as TTPs to sow political confusion in western liberal democracies, including the United States to an extent that the institutions of these democracies are undermined in terms of their public legitimacy (see Maurushat 2013). This comports with recent analysis within international relations that states such as Russia will pursue chaos implementation as a foreign policy strategy to counter the law and order established through 'thick' institutionalism and interdependence (Mousseau 2019, 196).

Russia's 2014 annexation of Crimea, its ongoing interference in Ukrainian political affairs for years, and its full invasion of Ukraine in early 2022, have led to strains in relations with the United States along with NATO more generally (NATO 2018, 20). Moreover, Russia used cyberattacks to recently target Estonia (a NATO member nation) (Štitilis et al. 2017). Jens Stoltenberg, (the NATO Secretary General) met with acting U.S. Defense Secretary Patrick Shanahan to discuss the ongoing threat of Russia to the Alliance (Shanahan 2019). This led to the cessation of cooperation and dialog between Russia and NATO (NATO 2018, 20). Once again, this demonstrated how Russia is quite skilled at using cyber to weaken the United States and its allies (and their diplomatic efforts) (NCS 2018, 2).

Russia mostly focuses on using cyber and by extension, information, as a way for it to exert influence in western nations (Weaver 2017). It has even engaged in misinformation and disinformation campaigns using graphic novels online to undermine democratic principles and governance (CISA Russia 2021a). Russia was also actively engaged in online activity to try to affect the outcome of the 2020 U.S. presidential election (CISA Russia 2021b). It engages in hybrid warfare and uses cyber to gain an advantage, especially when affording consideration to what occurred recently in Ukraine (Schwartz 2019). Russia adequately demonstrated its willingness to use hybrid TTPs to achieve a sense of synergy when coupling conventional strategies with military capabilities (Hoffman 2009; Reichborn-Kjennerud and Cullen 2016). In doing so, Russia was able to deflect attention away from what it was doing (Weaver and Johnson 2020).

Likewise, Russia integrates networks to figure out how to best implement mass disruption if and when required (GAO-19-204SP 2018, 3). More pointedly, it is involved in developing cyber as a weapon to provide itself with an edge over western nations that have greater superiority concerning military systems. Likewise, the information infrastructure and cyberspace networks of the U.S. have been vulnerable and thus becoming an easy target by state actors (including Russia) (Pomeroy 2019). At present, cyber activities by the adversaries of the United States have eroded the U.S. military advantages and, thus, negatively impacted U.S. economic prosperity (Pomeroy 2019).

Underscoring this, the former head of the Central Intelligence Agency (CIA), Gina Haspel when speaking at Auburn University about the threats that Russia poses, stated that this has garnered the attention of the agency in recent years (Haspel 2019). The CIA clearly understands Russia to be a major threat to the United States specifically, and western allies more broadly.

Moreover, Russia's state-sponsored cyber organizations have been developing complex software that can attack large companies for the extremely large amounts of money that they possess. Blockchain technology, for example, can inflict havoc in ways that were previously not seen (McDowell et al. 2019). The intelligence arm of the Russian military, the Main Intelligence Directorate (GRU), is one member of such a division that has been on the radar of the U.S. counterintelligence agencies. Two GRU officials were recently charged with interfering with the U.S. elections back in 2016 and were also found to be complicit with the NotPetya ransomware attack in 2017 (The Treasury Department 2018; Weaver and Johnson 2020). More surprisingly, the GRU officials were also found to be linked to using tradecraft and social engineering TTPs used by everyday cybercriminals (McDowell et al. 2019). The GRU was also named by the National Security Agency as a user of 'brute force' TTPs to gain access to public and private networks of the United States (NSA 2020). This further proves that hackers with a wide range of motives continue to target cyber infrastructure as their knowledge to acquire easy access to networks through phishing, its understanding of human nature, and how social media use can be exploited (Jewett 2018). Therefore, it is incumbent upon the United States and its allies to focus attention and efforts toward educating their public on cybersecurity, the threats thereof, and to raise public awareness.

The pursuit of the proliferation of cyber TTPs by adversaries and the exploitation of low-level cybersecurity awareness are not an understatement (McDowell et al. 2019). Kevin Mitnick, KnowBe4's Chief Hacking Officer, made a statement regarding his lack of surprise when he saw that Russians were using the same methods his company employs to test their clients' security (McDowell et al. 2019). Mitnick also stated that his engineers often realize success when employing simple techniques like what the Russians use, such as spear-phishing (Jewett 2018). Collectively, countries like the United States may be making modest progress toward achieving cybersecurity; however, the adversaries of the U.S. are constantly searching for weaknesses to spot and penetrate networks like one saw with the SolarWinds hack in 2020 (Weaver 2020). Russia was even called out by President Biden when he spoke to Putin about the uptick of Ransomware attacks instigated by Russia (White House Press Release 2021). Being the world's most vibrant economy, one could easily appreciate the difficulty with the U.S. being able to secure every network at all times (public and private). Through evolutionary steps moving beyond mass-mail phishing attempts, spear-phishing targets single, high-placed employees with bait emails; once an individual takes the bait, the hacking group is then subsequently able to access the entire network of an organization through their account causing destruction (Parmar 2012). The same TTPs, however, cannot be guarded against by using advanced firewalls and other security measures alone, because human error is what often opens the door to exposing networks to vulnerabilities.

In 2019, a recent former Chairman of the U.S. Joint Chiefs stated that Russia was involved not only in influencing elections in the United States but in activities directed at the allies of the U.S. as well (Dunford 2019). In recent years, twelve Russian GRU members were charged with their interference in the 2016 U.S. election. These officers used cyber TTPs while spear-phishing employees and volunteers of Hillary Clinton, a U.S. Presidential candidate, and the targets included the campaign's chairman (Weaver and Johnson 2020). Subsequently, these GRU members were able to hack into the computer networks of the Democratic National Committee (DNC) and the Democratic Congressional Campaign Committee (DCCC) and steal documents and emails, they covertly monitored the computer activity of dozens of employees, and modified hundreds of files of malicious computer code to illicitly acquire passwords and maintain access to these networks (McDowell et al. 2019). These officers then used the stolen information and released the material on websites such as DCLeaks.com and created fake Twitter and Facebook accounts. The GRU personnel originally claimed that they were "American hacktivists," and later professed to having exploited the computers and networks of state boards of elections, secretaries of state, and U.S. companies that supplied information related to voter data that was stored on specific computers (U.S. Department of Justice 2018).

The U.S. election meddling did not end there. Another group, the Internet Research Agency (IRA) LLC, was personally named in the 2016 U.S. election scandal (McDowell et al. 2019). The IRA, LLC is known for creating an abundance of fake online profiles by making use of legitimate citizens linked to real interest groups as well as those associated with local and state political parties on social media. The IRA later posted thousands of ads reaching millions of people all online with the specific intent of influencing vote casting. Through the acquisition of illegally acquired personally identifiable information from citizens of the U.S., the IRA was then able to open financial accounts, which further aided their activities (The Treasury Department 2018). Such incidents surfaced serious national security questions because the attack was so successful whereby the identity of those responsible was not understood until after the election was over.

In 2020, the United States saw similar attempts aimed at altering the election outcomes. The 2021 Worldwide Threat Assessment underscored Russian activities once again (WTA 2021, 11).

Like the WannaCry attack of 2017, NotPetya (another ransomware attack) saw more than 200,000 computers in 150 countries become infected with a malware known as ransomware; it locked users out and would not allow them to operate unless they paid a ransom or infected at least two other people with the same virus (McDowell et al. 2019). An example occurred when NotPetya devastated the United Kingdom's healthcare system, resulting in the cancellation of hundreds of surgical procedures due to the hospitals' computers being infected with the malware (Jasper 2017).

Russia, just like many of the other state actors covered in this book, sees the reliance on space by the United States as a vulnerability. Russia, as a result of possessing this knowledge, is developing military doctrine to modernize its warfare capabilities while it also looks to use space-based systems and cyber to monitor, track, and target U.S. (and, by extension, allied) forces (Security in Space 2019). It is investing significant monetary resources to do so, but to a lesser extent than China with the express intent of denying the United States and allies from accessing information from space-based platforms (Security in Space 2019, 23). Of the over 130 satellites Russia has in orbit, more than half are dedicated to its communications network and roughly 18% are used for ISR collection; another 18% have commercial applications (Security in Space 2019).

Accordingly, Russia most likely will continue to invest in its space-based programs to enhance its ISR capabilities and to improve its communications infrastructure (Security in Space 2019, 33). Its other programs also include such things as satellite communications and the Global Navigation Satellite System or GLONASS (Security in Space 2019, 27). Likewise, Russia is investing in counter-space capabilities while it looks to electronic warfare technology to minimize the technical edges that countries like the United States have in the cyber domain (SFTR1 2018; SFTR2 2018).

Historically, Russia has been quite apprehensive of the North Atlantic Treaty Organization (NATO) and presently sees NATO as its main culprit and by extension the United States (Putin 2014). To help achieve success, Russia is frequently turning to cyber TTPs as a way to help ensure early gains in the initial stages of a future conflict (Chekinov and Bogdanov 2012). Moreover, its leadership views the information sphere as a key domain in military conflicts (Chekinov and Bogdanov 2013). This country's military doctrine attests to the value of electronic warfare and by extension the use of cyber means (Putin 2014). Putin (2014) underscored this through his recent military doctrine by ascribing importance to the creation of such capabilities.

The intent behind Russia's actions will be to control all facets of the information environment (Russia Military Power 2017, 38). Russia's end state would be for it to see the disabling of adversarial computers and networks in addition to the disruption of the electromagnetic spectrum (Donskov and Nikitin 2005). Accordingly, Russia's aspirations look to include the development of its accoutrement of full-spectrum electronic warfare capabilities to hedge its position in relation to NATO C4ISR systems (Russia Military Power 2017, 42).

What is most troubling about Russia centers on the U.S. Intelligence Community's assessment that it is staging cyber-attacks and influence operations targeted at the United States with the goal being able to influence U.S. politics (WTA 2019, 5; 2021, 11). While it is doing this, Russia's intelligence and security services are adamantly targeting information systems of not only the United States, but those of NATO, and the Five Eyes partners as well (WTA 2019). An example showcasing Russia's adeptness at cyber-attacks was demonstrated when it recently disrupted the United States' electrical power distribution network (WTA 2019).

The former director of the NSA, Admiral Rogers, noted that Russia's cyber and information influences go beyond elections and include frequent attacks on U.S. private and public sector organizations as well (Rogers 2017). Russian activities in cyber espionage have harmed U.S. and international businesses and have not been effectively thwarted from continuing their practices (NCS 2018, 1). The Intelligence Community assesses it as a major cyber player on the world stage (WTA 2017, 1; 2021, 11). Like China, it has invested in CNA and CNE capabilities to steal secrets from other countries to gain advantages (Weaver 2017, 10). Likewise, it is gravely concerned about cybersecurity (CND) to protect itself from exploitation by other state and non-state actors (Crosston 2016, 124).

While Russia engages in cyber espionage, it is enhancing its computer network defenses; this was explicitly stated in a Russian cyber defense plan that dates to December 2016 (Russia Plan 2016). Russia most likely wants to advance its position diplomatically. Further discussion with a recent former head of the NSA of the United States revealed that this has resulted in the loss of billions of dollars and even state secrets (Rogers 2017). Bill Priestap, when he testified before the U.S. Senate Intelligence Committee, stated that Russia knows that it can't confront the U.S. directly with its military and is investing in cyber to weaken this nation and its allies (CSPAN 2017). When shifting to a NATO ally, Mr. Hjort Frederiksen, Denmark's Defense Minister stated, "This is part of a continuing war from the Russian side in this field, where we are seeing a very aggressive Russia" in response to Russia's attack on its military networks (MacFarquhar 2017). Even though Russia was not successful at accessing classified information, the actions were still troubling. The U.S. Intelligence Community has also assessed that Russia will most likely continue to use cyber to implement information operations to affect diplomatic initiatives and military operations around the world (WTA 2017, 1). Russia's cyber use as an espionage tool has helped leverage its asymmetric military capabilities while trying to find vulnerabilities in U.S military systems (Weaver 2017). Other evidence emerging from secondary sources shows the adeptness of Russia particularly with its competence at using cyber espionage to more effectively infiltrate government and private sector networks throughout the United States (Weaver 2017). Russia will continue to invest in cyber TTPs to gain a better position against western technology through exploiting weaknesses in western industrial bases, public & private networks, and military capabilities (Weaver 2019).

Russia has been a prominent player on the global stage in recent years. More pointedly, it has exerted significant influence in Eastern Europe to the concern of NATO, and by extension—the United States as alliance member nations. When revisiting the YIRTM-C (M) and by examining the evidence in Annex 9.1 interesting results surface.

Russia has overall used cyber to help advance (while harming others) using the information and military components of the model. Like many of the other state and non-state actors analyzed in this book, Russia knows that it can't confront the United States militarily with NATO and win. Accordingly, it knows it must invest in cyber units to look for vulnerability in western nations, their networks, infrastructure, and tech-laden military weapon systems.

Russia has also turned to cyber to exert economic influence. It used cyber to conduct espionage to gain access to information to improve its financial position while having an impact on the economic prosperity of nations like the United States.

To an extent equal to its use of cyber through the economic instrument, Russia has been quite successful and continues to use cyber conveyance to enhance diplomatic influence to present its image as a strong and resilient actor. Likewise, it has silhouetted itself as a viable partner to counter western influence from the likes of the United States to non-western nations including Syria, Iran, and even Turkey through the sale of advanced air defense systems (Turkey, who is a NATO member).

Annex 9.1: Russia

How/Why	D.I.M.E	Source type	Author	Date	Page(s)
Russia is investing in counter-space capabilities and looks to electronic warfare technology to minimize the United States' technical edges	I, M	Testimony	SFTR1, SFTR2	2018	NA

(continued)

(continued)

How/Why	D.I.M.E	Source type	Author	Date	Page(s)
Russia is developing its military doctrine while it modernizes its warfare capabilities while looking to use space-based systems and cyber to monitor, track, and target U.S. and allied forces	I, M	Plans and Systems	Security in Space	2019	iii
It is investing extensively in cyber but to a lesser extent than China to deny the United States and allies from accessing information from space-based platforms	I, M	Government Document	Security in Space	2019	24
More than half of its 130 satellites are dedicated to Russia's communications network and roughly 18% are used for ISR collection; roughly 18% are also used for commercial purposes	I, M, E	Government Document	Security in Space	2019	27
Russia identifies NATO as its main threat and by extension the United States	M	Plans and Systems	Putin	2014	NP
Russia is investing in and turning to cyber TTPs to help ensure quick gains in the initial stages of a future conflict	I, M, E	Journal	Chekinov and Bogdanov	2012	NA
Russia considers the information sphere as a key domain in military conflicts	I, M	Journal	Chekinov and Bogdanov	2013	NA
Russia's goals include the development of its own array of full spectrum electronic warfare capabilities to hedge against NATO C4ISR systems	I, M	Government Document	Russia Military Power	2017	42

(continued)

(continued)

How/Why	D.I.M.E	Source type	Author	Date	Page(s)
In its military doctrine, Russia attests to the value of electronic warfare and by extension the use of cyber means	I, M	Plans and Systems	Putin	2014	NP
Putin underscores cyber's significance in his military doctrine by ascribing importance to the creation of such capabilities	I, M	Plans and Systems	Putin	2014	NP
The U.S. Intelligence Community assesses that Russia is staging cyber-attacks and influence operations targeted at the United States	D, I, M, E	Government Document	WTA	2018	5
Russian intelligence and security services are aggressively targeting information systems not only of the United States, but those of NATO, and the Five Eyes partners as well	D, I, M	Government Document	WTA	2018	5
Russia demonstrated adeptness at cyber-attacks as was realized when it disrupted the United States' electrical power distribution network in recent times	I, E	Government Document	WTA	2018	6
Russia mostly focused on using cyber and information as a way to conduct influence in western nations	I	Book Chapter	Weaver; WTA	2017; 2021	29–54; 11
Russia's information and cyber move beyond elections and include frequent attacks on U.S. private and public sector organizations in discussion that occurred with the director of the NSA	D, I	Interview	Rogers	2017	NA

(continued)

(continued)

How/Why	D.I.M.E	Source type	Author	Date	Page(s)
Russia is enhancing its computer network defenses; this was stated in a Russian cyber defense plan	D, I, M	Plans and Systems	Russia Plan	2016	NA
Russia is investing in cyber as a way to weaken this nation and its allies	I	Interview	CSPAN	2017	NA
Other evidence from secondary sources shows Russia's adeptness particularly with its competence at using cyber espionage as a way to infiltrate into government and private sector networks throughout the United States	I, M, E	Book Chapter	Weaver	2017	29–54
Russia will continue to invest in cyber TTPs to flex against western technology through exploiting weaknesses in industrial bases, western public & private networks, and military capabilities	I, M, E	Book	Weaver	2019	29–54
Gina Haspel, who headed the CIA, spoke at Auburn University about the threats that Russia poses and how this has garnered the attention of the agency in recent years	D, I, M, E	Speech	Haspel	2019	NP
U.S. information infrastructure and cyberspace networks have been vulnerable and increasingly become an easy target by state actors including Russia	D, I, M, E	Book Chapter	Pomeroy	2019	320–322
Cyber activities by the adversaries of the U.S. have eroded the U.S. military advantages and negatively impacted the U.S. economic prosperity	M, E	Book Chapter	Pomeroy	2019	320–322

(continued)

(continued)

How/Why	D.I.M.E	Source type	Author	Date	Page(s)
Russia uses blockchain technology and has the capability to inflict havoc	D, I, M, E	Book Chapter	McDowell, Walker, and Meyers	2019	293
Two GRU officials were charged with interfering with the U.S. elections back in 2016; they were also found to be involved with the NotPetya attack in 2017	D, I	Book Chapter	McDowell, Walker, and Meyers	2019	293
Twelve Russian GRU members were charged with the 2016 U.S. election interference	D, I	Book Chapter	McDowell, Walker, and Meyers	2019	310
In March 2019, a recent former Chairman of the U.S. Joint Chiefs stated that Russia was involved not only in influencing elections in the United States, but in activities directed at U.S. allies as well	D, I	Interview	Dunford	2019	NP
Russia is quite adept in using cyber to weaken the United States and its allies (and their diplomatic efforts)	D	Government Document	NCS	2018	2
Russia's cyber espionage activities have harmed U.S. and international businesses and have not been thwarted from continuing their practices	E	Government Document	NCS	2018	1
Russia used online activity to try to affect the outcome of the 2020 U.S. presidential election	D	Interview	CISA Russia	2021b	NP
Russia was also actively engaged in online activity to try to affect the outcome of the 2020 U.S. presidential election	I, E	Government Document	NSA	2020	NP
There is an uptick of Ransomware attacks instigated by Russia	E	Press Release	White House Press Release	2021	NP

REFERENCES

Boghardt, T. 2009. Soviet Bloc Intelligence and Its AIDS Disinformation Campaign. *Studies in Intelligence* 53 (4): 1–24.

Chekinov, Sergey G., and Sergey A. Bogdanov. 2012. Initial Periods of Wars and Their Impacts on a Country's Preparations for a Future War. *Military Thought*, December 31.

Chekinov, Sergey G., and Sergey A. Bogdanov. 2013. Asymmetric Actions to Maintain Russia's Military Security. *Military Thought*, June 20.

Chesney, R., and D. Citron. 2019. Deepfakes and the New Disinformation War: The Coming Age of Post-Truth Geopolitics. *Foreign Affairs* 98: 147.

CIA. 2019. The World Factbook, Russia. https://www.cia.gov/library/publications/resources/the-world-factbook/geos/rs.html. Accessed 17 June 2019.

CISA Russia. 2021a. Resilience Series Graphic Novels. https://www.cisa.gov/resilience-series-graphic-novels. Accessed 28 July 2021.

CISA Russia. 2021b. Statement from CISA Director Krebs on Election Security Announcement. https://www.cisa.gov/news/2020/10/21/statement-cisa-director-krebs-election-security-announcement. Accessed 28 July 2021.

Crosston, Matthew. 2016. Bringing Non-Western Cultures and Conditions into Comparative Intelligence Perspectives: India, Russia, and China. *International Journal of Intelligence and Counterintelligence* 29 (1): 110–131.

CSPAN. 2017. *Russian Interference in U.S.* Elections. https://www.c-span.org/video/?430128-1/senate-intel-panel-told-21-states-targeted-russia-2016-election. Accessed 23 July 2017.

Donskov, Colonel Yu Ye, and Lieutenant Colonel O.G. Nikitin. 2005. Special Information Operations in Armed Conflict. *Military Thought*, September 30.

Dunford, Joe. 2019. Dunford Describes U.S. Great Power Competition with Russia, China. https://dod.defense.gov/News/Article/Article/1791811/dunford-describes-us-great-power-competition-with-russia-china/. Accessed 21 July 2019.

GAO-19-204SP. 2018. Report to Congressional Committees National Security Long-Range Emerging Threats Facing the United States as Identified by Federal Agencies. https://www.gao.gov/assets/700/695981.pdf. Accessed 23 July 2019.

Haspel, Gina. 2019. CIA Director Gina Haspel Speaks at Auburn University. https://www.cia.gov/news-information/speeches-testimony/2019-speeches-testimony/dcia-haspel-auburn-university-speech.html. Accessed 17 June 2019.

Hoffman, F.G. 2009. *Hybrid Threats: Reconceptualizing the Evolving Character of Modern Conflict*. Washington, DC: Institute for National Strategic Studies, National Defense University.

INSSG. 2021. Interim National Security Strategic Guidance. https://www.whi
tehouse.gov/wp-content/uploads/2021/03/NSC-1v2.pdf. Accessed 17 Aug
2021.

Jasper, Scott. 2017. *Russia and Ransomware: Stop the Act, Not the Actor.*
Calhoun: The NPS Institutional Archive Space Repository.

Jewett, Jennifer. 2018. *The Business JournalsKnowBe4 analysis: Lack of Security
Awareness Training Allowed Russians to Hack American Election*, July 17.
Accessed 23 Oct 2018. https://www.bizjournals.com/prnewswire/press_rel
eases/2018/07/17/FL55419.

MacFarquhar, Neil. 2017. *Denmark Says 'Key Elements' of Russian Government
Hacked Defense Ministry.* https://www.nytimes.com/2017/04/24/world/
europe/russia-denmark-hacking-cyberattack-defense-ministry.html. Accessed
27 July 2017.

Maurushat, A. 2013. From Cybercrime to Cyberwar: Security Through Obscurity or Security Through Absurdity? *Canadian Foreign Policy Journal* 19 (2):
119–122.

McDowell, Nathan, Ethan Walker, and Matthew Meyers. 2019. Prominent
Cybersecurity Issues for the United States: A Qualitative Assessment. In *Global
Intelligence Priorities (from the Perspective of the United States).* Nova Science
Publishers.

Mousseau, M. 2019. The End of War: How a Robust Marketplace and Liberal
Hegemony Are Leading to Perpetual World Peace. *International Security* 44
(1): 160–196.

NATO. 2018. *The Secretary General's Annual Report 2018.* https://www.nato.
int/cps/en/natohq/topics_164559.htm#:~:text=On%2014%20March%202
019%2C%20Secretary%20General%20Jens%20Stoltenberg,to%20new%20chal
lenges%2C%20and%20. Accessed 11 June 2019.

NCS. 2018. National Cyber Strategy of the United States of America. https://
www.whitehouse.gov/wp-content/uploads/2018/09/National-Cyber-Str
ategy.pdf. Accessed 22 July 2019 and Prosperity in the Digital Age. https://
www.securitepublique.gc.ca/cnt/rsrcs/pblctns/ntnl-cbr-scrt-strtg/ntnl-cbr-
scrt-strtg-en.pdf. Accessed 23 July 2019.

NSA. 2020. Russian GRU Conducting Global Brute Force Campaign to
Compromise Enterprise and Cloud Environments. https://www.nsa.gov/
news-features/press-room/Article/2677750/nsa-partners-release-cybersecu
rity-advisory-on-brute-force-global-cyber-campaign/. Accessed 6 July 2020.

Parmar Bimal, Faronics. 2012. Protecting Against Spear-Phishing. *Computer
Fraud & Security.* https://www.faronics.com/assets/CFS_2012-01_Jan.pdf.
Accessed 23 Oct 2018.

Pomeroy, Jennifer Yongmei. 2019. Challenges of the U.S. National Security and
Moving Forward. In *Global Intelligence Priorities (from the Perspective of the
United States).* Nova Science Publishers.

Putin, Vladimir. 2014. *Military Doctrine of the Russian Federation*. https://www.offiziere.ch/wp-content/uploads-001/2015/08/Russia-s-2014-Military-Doctrine.pdf. Accessed 10 June 2019.

Reichborn-Kjennerud, E., and P. Cullen. 2016. *What Is Hybrid Warfare?* Norwegian Institute of International Affairs. Policy Brief, 1, 2016.

Rogers, Michael. 2017. Admiral Michael S. Rogers (USN), Director, National Security Agency, and Commander, U.S. Cyber Command, Delivers Remarks at The New America Foundation Conference on CYBERSECURITY. https://www.nsa.gov/news-features/speeches-testimonies/speeches/022315-new-america-foundation.shtml. Accessed 20 July 2017.

Russia Military Power. 2017. Russia Military Power Building a Military to Support Great Power Aspirations. https://www.dia.mil/Portals/27/Documents/News/Military%20Power%20Publications/Russia%20Military%20Power%20Report%202017.pdf?ver=2017-06-28-144235-937. Accessed 11 June 2019.

Russia Plan. 2016. http://www.ieee.es/Galerias/fichero/OtrasPublicaciones/Internacional/2016/Russian-National-Security-Strategy-31Dec2015.pdf. Accessed 17 Dec 2019.

Schwartz, Tamara B. 2019. A Dynamic Cyber-Based View of the Firm. Dissertation, Temple University.

Security in Space. 2019. *Challenges to Security in Space*. https://www.dia.mil/Portals/27/Documents/News/Military%20Power%20Publications/Space_Threat_V14_020119_sm.pdf. Accessed 10 June 2019.

SFTR1. 2018. Statement for the Record: Worldwide Threat Assessment of the U.S. Intelligence Community. Office of the Director of National Intelligence, 13 February 2018.

SFTR2. 2018. Statement for the Record: Worldwide Threat Assessment of the U.S. Intelligence Community. Office of the Director of National Intelligence, 13 February 2018.

Shanahan, Patrick. 2019. Readout of Acting Secretary of Defense Patrick M. Shanahan's Meeting with NATO Secretary General Jens Stoltenberg, News Release No: NR-016-19. https://dod.defense.gov/News/News-Releases/News-Release-View/Article/1742285/readout-of-acting-secretary-of-defense-patrick-m-shanahans-meeting-with-nato-se/. Accessed 21 July 2019.

Štitilis, Darius, Paulius Pakutinskasb, and Inga Malinauskait. 2017. EU and NATO Cybersecurity Strategies and National Cyber Strategies: A Comparative Analysis. *Security Journal* 30 (4): 1151–1168.

The Treasury Department. 2018. *Treasury Sanctions Russian Cyber Actors for Interference with the 2016 U.S. Elections and Malicious Cyber-att$acks*. https://home.treasury.gov/news/press-releases/sm0312. Accessed 23 Sept 2018.

UK. 2021. UK and Allies Hold Chinese State Responsible for a Pervasive Pattern of Hacking. https://www.gov.uk/government/news/uk-and-allies-hold-chinese-state-responsible-for-a-pervasive-pattern-of-hacking. Accessed 20 July 2021.

U.S. Department of Justice. 2018. *Grand Jury Indicts 12 Russian Intelligence Officers for Hacking Offenses Related to the 2016 Election.* Office of Public Affairs. https://www.justice.gov/opa/pr/grand-jury-indicts-12-russian-intelligence-officers-hacking-offenses-related-2016-election. Accessed 23 Sept 2018.

Weaver, John M. 2017. Cyber Threats to the National Security of the United States: A Qualitative Assessment. In *Focus on Terrorism (Volume 15).* Nova Science Publishers.

Weaver, John M. 2019. *United Nations Security Council Permanent Member Perspectives Implications for U.S. and Global Intelligence Professionals.* Peter Lang Publishing.

Weaver, John M. 2020. Hackers Are a Major Threat to National Security, and It Will Only Get Worse. *Pennlive.* Pennsylvania, USA.

Weaver, John M., and Benjamin Johnson. 2020. *Cyber Security Challenges Confronting Canada and the United States.* New York, USA: Peter Lang Publishing.

White House Press Release. 2021. Background Press Call by Senior Administration Officials on President Biden's Call with President Putin on Russia. https://www.whitehouse.gov/briefing-room/speeches-remarks/2021/07/09/background-press-call-by-senior-administration-officials-on-president-bidens-call-with-president-putin-of-russia/. Accessed 14 July 2021.

WTA. 2017. *Worldwide Threat Assessment.* https://www.dni.gov/files/documents/Newsroom/Testimonies/SSCI%20Unclassified%20SFR%20-%20Final.pdf. Accessed 20 Jan 2018.

WTA. 2019. *Worldwide Threat Assessment of the US Intelligence Community.* https://completethreatpreparedness.com/wp-content/uploads/2019/02/2019-ODNI-Worldwide-Threat-Assessment.pdf. Accessed 11 June 2019.

WTA. 2021. Worldwide Threat Assessment of the US Intelligence Community. https://www.dni.gov/files/ODNI/documents/assessments/ATA-2021-Unclassified-Report.pdf. Accessed 6 July 2021.

CHAPTER 10

Analysis, Findings, Assessment

Abstract This chapter collectively looks at the instruments of national power through the lens of the York Intelligence Red Team Model-Cyber (modified) and collectively answers the research questions to see holistically, how and why the actors are using the instruments in this model through cyber operations to weaken the United States.

Keywords YIRTM-C (M) · C4ISR · Cyber

The focus of this chapter examines the YIRTM-C (M) and its application to demonstrate how state and non-state actors are making use of the instruments of power against the United States. The analysis follows in the paragraphs below.

Presidential Policy Directive 41 (PPD-41) was implemented in 2016 (Obama 2016). PPD-41, also known as the United States Cyber Incident Coordination, was an acknowledgment by the executive branch regarding its understanding of the vulnerabilities and weaknesses of public and private sector organizations to various malicious activities, malfunction, and other cyber activities, and substantiates a need for concerted governmental effort pertaining to cyber incidents.

J. M. Weaver, *The U.S. Cybersecurity and Intelligence Analysis*, https://doi.org/10.1007/978-3-030-95841-1_10

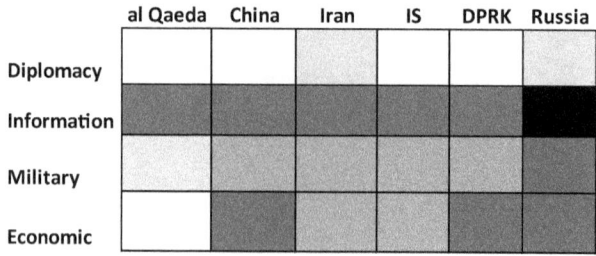

Fig. 10.1 YIRTM-C (M) matrix

When circling back to the mid-2000s, cyber incidents have burgeoned markedly in every year since 2006 through 2014 by a 12-fold factor (GAO 15-758T 2015, 7). These incidents have included the likes of unauthorized access, improper usage, suspicious network activity, scans/probes/and attempted access, and malicious code injection, among others. Moreover, these encapsulated over 65,000 incidents on a per annum basis and continue to remain on an upward trajectory. Dating back to 2014, 17 out of 24 major U.S federal agencies reported significant cyber incidents (GAO 2015). Though this study focused solely on secondary data sources, the researcher was able to concentrate his findings over a seven-year temporal period to see just what was taking place.

Governments often make use of vetted hackers and cyber companies to check for vulnerabilities and this is common in North America. Even the U.S. Defense Department (DOD), as acknowledged through a press release in early March 2016, had understood the threat and invited cyber experts to openly attempt the hacking of the DOD's network to test its cyber defensive capabilities (Cook 2016). The intention was to identify vulnerabilities in its unclassified network. Overall, the Secretary of Defense wanted to build resilience into the cyber infrastructure while pushing for innovative solutions to enhance the department's cyber protection. Attacks on the United States are becoming much more destructive and disruptive as seen in recent years (GAO-16-332 2016).

Once again, the author balanced information pursuant to the triangulation model displayed in Fig. 2.2 as the data applied to the YIRTM-C (M) represented in Fig. 2.1. That stated, the YIRTM-C (M) proved to be useful. Figure 10.1 depicts the breakout by instrument (variable) and where most of the impact was utilized by the actors regarding the individual instruments as implemented by each actor. The darker the color, the more reliant the actor was on the instrument to exert pressure countering U.S influence in the world.

After affording consideration to the data in general and the instruments of national power more specifically, several foes (countries and non-state actors) emerge. The country demonstrating the most effectiveness at using cyber to reduce the United States' power is Russia. China and Iran are also notable actors. All countries have robust military capabilities, and Russia and China are both pursuing out-of-area operations to demonstrate their strength to the world. Moreover, the evidence reflected in the chapter annexes points to Russia and China as being quite adept at using cyber espionage to infiltrate private sector and government networks throughout the United States. According to Rudner (2013, 462), several reports further showed that China was engaged in state sponsored CNO through its targeting of 72 organizations around the globe which included among others, the United Nations, governments, and corporations. Through careful application of the variables in the YIRTM-C (M), the nations identified in this book seemingly are committed to using the information and economic capabilities that can be gained from gathering information. That stated, both Russia and China are making use of technology that will reduce the conventional weapons' superiority currently possessed by other nations, including the United States. More specifically, both China and Russia are actively pursuing the acquisition of scientific data that could reduce the stealth advantages held by the United States' military. Other indicators point to Chinese and Russian agents being successful in specifically using the information, military, and economic components of the YIRTM-C (M), and are less interested in the diplomacy component, though many countries are open to debate the imposition of limitations on how nation-states intend to employ cyber.

Both Iran and North Korea have been beleaguered by sanctions in recent years and experienced significant economic hardship. Though sanctions have affected their nations militarily and the wellbeing of their populations, both Iran and North Korea have been able to exert influence in the Middle Eastern and the Indo-Pacific areas respectively. To help further their positions, both have sought cyber capabilities and have seen these as ways to implement asymmetric TTPs to project power regionally. Indicators point to North Korea and Iran using cyber capabilities to demonstrate the economic impact each can have on the commercial sector and as a way to achieve political points to promote their diplomatic objectives. According to Rudner (2013, 460), Iran demonstrated this proclivity through its implementation of the 'Iranian Cyber Army' and has most likely launched CNO against the U.S government and business websites according to data used in this research. These two countries have clearly demonstrated their adeptness for using the diplomatic, informational, and economic components of the YITRM-C (M) through cyber to advance their positions while trying to weaken the stature of western nations like the United States.

The Islamic State has made significant forays into using cyber to advance its position. Most notably, it has significantly planned, organized, and executed a cyber campaign to more aptly promote the organization's ideology and to recruit new operatives to the cause. This terror group has demonstrated a penchant for social media use to the extent that it is willing to allocate its most talented members to its cyber activities. Information is often akin to power and ISIL recognizes that by using tech-savvy operatives who can help it articulate its message through professional messages and videos transmitted via the internet and social media can resonate favorably in the minds of young impressionable adults and offers a significant and viable tool in bringing talent into its ranks. It is also through social media's use that the Islamic State has been able to generate support among other belligerent terror organizations like Boko Haram and others helping to coalesce other like-minded terror organizations to its cause.

IS, through its effort to promote its information while also surviving militarily, has published its own cybersecurity survival guide (ISIS 2015). In this plan, it has promoted the use of TTPs such as disabling mobile phone location services, encryption, ensuring operational security when taking photos, how to operate in internet denied environments, the use of a Russian encrypted messaging application called 'Telegram,' as well promoting instruction on how to browse anonymously online.

When circling back to the YIRTM-C (M), the Islamic State has shown success in using cyber conveyance for diplomacy and information propagation. Ashton Carter, a former U.S. Secretary of Defense, even underscored the importance of the Islamic State's use of cyber when testifying before the Senate Armed Services Committee in late 2015 (Carter 2015). At the time, he informed the senators of the Defense Department's need to implement more effective strategic courses of action to overcome cyber threats emanating from this group. The following month,

Carter went farther when in discussion with allies at the Ecole Militaire in Paris, France (Carter1 2016). During his speech, Carter specifically referenced the topic of the cyber defense activities of allied nations and their efforts in using cyber to prosecute the fight against IS. Even a recent former head of the United States Special Operations Command understood the relevance and strategic importance of countering propaganda coming out of terror organizations. In testimony before the House Armed Services Committee in March 2016, he informed lawmakers of the necessity to prevent the spread of terror ideology online (Shane 2016).

Al Qaeda (and its affiliates) has not been as effective as the other actors in this book at using the YIRTM-C (M) to project power. Nonetheless, the terrorist organization intends to utilize the information component the most and, to a lesser extent, the military instrument in coordination with cyber to threaten the economic infrastructure of adversarial countries via what it refers to as 'Economic Jihad' (Rudner 2013, 455). Moreover, to amplify this, Jihadists conducted a very successful cyber-attack against an American railway company (Runder 2013, 456). This has shown al Qaeda's advancement in cyber operations in recent years.

Through studying the primary actors bent on using cyber and after reading into the TTPs used by non-state and state actors alike and after thoroughly analyzing the data, this study finds that most are interested in using cyber as a way to incorporate both the information and the military instruments of power as the primary pillars. Moreover, the data from the chapter annexes show that these are the most frequented components that have resonated throughout secondary data sources. Actors seeking employment of cyber see the advantages that it offers in achieving desired outcomes/effects by leveraging information technology to enhance their standing and position, thereby balancing to greater or lesser degrees the asymmetry experienced between these specific actors and the United States in terms of conventional forces and power, especially the information and military components of the YIRTM-C (M).

The economic instrument of power showed prominence but to a degree lesser than the information and military instruments of power. Those most likely to use cyber for military purposes require mature systems, a cadre of capable hackers, and an ability to mask their intentions to bring success to fruition. The two most notable actors that are in possession of this capability are China and Russia. Chakarova and

Kokcharov (2016) write about the alleged implementation of Russian cyber-attacks in Ukraine resulting in a power outage affecting over 100,000 customers late in 2015 (to exert economic impact). An assessment of China's People's Liberation Army capability also showed they had the ability to acquire systems and capabilities to be used for cyber-attacks (Regional Focus 2015). Cities in the United States recently have seen attempts made on their power grids which could also cause economic harm like what one saw in Ukraine.

The military of the United States (and recent success) have relied on technological innovation, and because of this, exposure to exploitation and attacks can make its military vulnerable to cyber threats. C4ISR system compromise could negate the advantages its military possesses over conventional military threats coming from China and Russia. Admiral Michael Rogers, the former head of the U.S. Cyber Command informed the Senate Armed Services Committee about his concerns and how they gravitate around the ability of China and Russia (and to a lesser extent Iran), and the cybersecurity concerns that they present against critical infrastructure of nation-states (Associated Press 2016). Arguably, one could make the argument that the United States can counter with CNA, but whoever stands to gain the most might be an actor who has either adequately implemented cyber defensive measures or has non-C4ISR dependencies to allow it to continue moving forward with military operations absent digitally integrated networks and sophisticated communication systems.

When considering the YIRTM-C (M), the one instrument that was least considered as a way to leverage one's position against the United States was diplomacy. Due to the asymmetric nature of CNO, one can quickly grasp that it is not in a lesser power's benefit to see the realization of formalized cyber agreements. Further, when considering state and non-state actors, many realize they cannot compete equally with the United States economically nor can they confront their militaries head-on with any chances of sustainable success.

The first research question looked at how state and non-state actors are using CNO in the context of the YIRTM-C (M) to weaken the United States' position in the world. When looking at this question, one must consider TTPs used against them. Specifically, these included unauthorized access, engagement in suspicious network activity, scans/probes/and attempted access, and malicious code insertion into networks and systems. Importantly, these CNOs have increased significantly in recent times. According to the secondary sources, China and Russia seemingly have been engaged in cyber espionage activities the world over and have found continued success in using them, suggesting that they will continue to employ such activities.

Smaller nation-states have acknowledged the viability in CNO and have enhanced these asymmetric capabilities. Both North Korea and Iran have furthered their capabilities in the field of cyber to exert an impact on the United States' commercial sector. Iran's militarization of cyber has been quite influential in targeting business and government networks, and websites alike according to secondary data sources.

Similarly, non-state actors like IS have made use of cyber to their advantage. The Islamic State has effectively made use of its cyber campaign to expand its ranks through recruiting new operatives sympathetic to its cause. IS' tech-savvy personnel have successfully exploited social media through well-narrated videos and messages seeking to turn impressionable men (and now women) to coalesce around its cause. Indeed, evidence suggests that ISIL has been successfully targeting well-educated young men for recruitment, especially those skilled in STEM (Rose 2015; Jaafari 2015).

Data reflected in chapter annexes also show the disaggregation by source and instrument to demonstrate more aptly what is taking place. More pointedly, the evidence has shown that over time, the level of sophistication of attacks has increased in recent years. Likewise, data interpretation considers the multiplicity of targets ranging from attempts on the power grid, the commercial sector, and critical North American infrastructure, all while the U.S moves forward with enhancements of its cyber capabilities.

The second question investigated why actors were making use of cyber as a way to weaken the leadership positions of the United States. Once again, when turning to chapter annexes, one will see that organization and state actors are using CNO as a way to perform 'economic jihad' against countries aligned with values akin to those of the United States. Likewise, potential adversaries are using cyber to gain access to trade secrets thereby reducing costs associated with their research and development (R&D) efforts and this applies not only to commercially available information technology, but to weapon systems as well. Lending support to this has been manifested through greater investment in the cyber capabilities of state and non-state actors alike.

Most state and non-state actors realize that they cannot confront the United States directly through conventional methods and win. The United States is extremely wealthy and arguably has the most potent and capable military in the world. Thus, leveraging cyber to achieve asymmetric advantages offers the best chances for these actors to be successful at weakening this country. This further helps underscores why these actors are using cyber TTPs to their advantage since this affords both state and non-state actors with standoff (to avoid a head-to-head engagement) and it helps them achieve anonymity, thereby granting the benefits of classic espionage through 'grey zone' conflict.

Why is cyber a concern for the United States? First, there are defense agreements in place. The most important includes NATO.

Secondly, the viability of this nation is tied to cybersecurity and how well it responds to cyber threats. More pointedly, as was covered earlier, the United States works in partnerships at, within, and beyond its borders to enhance security and to foster the legitimate flow of people, goods, and services (DOS 2019).

After disaggregating the data based on the application the two models used in this book, the assessment showed government documents, legislation, and policy underscore the use of mostly the "I," "M," and "E" power components to weaken the U.S longitudinally. Moreover, the press releases, testimony, and other oral accounts also underpin the use of the "I" and "E" instruments and to a slightly lesser extent, the use of "M." Finally, when looking at plans and systems, these underscore the viability of the use of the "I" and "M" instruments.

REFERENCES

Associated Press. 2016. *Russia, China Are Greatest Cyberthreats, but Iran Is Growing*. http://www.nytimes.com/aponline/2016/04/05/us/politics/ap-us-military-cyberwar.html?_r=1. Accessed 11 Apr 2016.

Carter, Ashton. 2015. *Statement on the Counter-ISIL Campaign Before the Senate Armed Services Committee*. http://www.defense.gov/News/Speeches/Speech-View/Article/633510/statement-on-the-counter-isil-campaign-before-the-senate-armed-services-committ. Accessed 26 Feb 2016.

Carter1, Ashton. 2016. *Counter ISIL Campaign Remarks at the Ecole Militarire, Paris*. http://www.defense.gov/News/Speeches/Speech-View/Article/643904/counter-isil-campaign-remarks-at-the-ecole-militaire-paris. Accessed 26 Feb 2016.

Chakarova, Lora, and Alex Kokcharov. 2016. Critical European Infrastructure Increasingly Likely to Be Targeted by Russian Cyber Groups in Three-Year Outlook. *Janes Defence*, February 10, 2015.

Cook, Peter. 2016. Statement by Pentagon Press Secretary Peter Cook on DoD's "Hack the Pentagon" Cybersecurity Initiative. http://www.defense. gov/News/News-Releases/News-Release-View/Article/684106/statement-by-pentagon-press-secretary-peter-cook-on-dods-hack-the-pentagon-cybe. Accessed 6 Mar 2016.

DOS. 2019. *U.S. Relations with Canada*. https://www.state.gov/u-s-relations-with-canada/. Accessed 9 June 2019.

GAO. 2015. *Cybersecurity*. http://www.gao.gov/key_issues/cybersecurity/iss uesummary. Accessed 18 Feb 2016.

GAO-15-758T. 2015. *Information Security Cyber Threats and Data Breaches Illustrate Need for Stronger Controls across Federal Agencies*. http://www.gao. gov/assets/680/671253.pdf. Accessed 16 Feb 2016.

GAO-16-332. 2016. *Civil Support DOD Needs to Clarify Its Roles and Responsibilities for Defense Support of Civil Authorities, During Cyber Incidents*. http://www.gao.gov/assets/680/676322.pdf. Accessed 6 Apr 2016.

ISIS. 2015. *Several Cybersecurity to Protect Your Account in the Social*. https://www.wired.com/wp-content/uploads/2015/11/ISIS-OPSEC-Guide.pdf. Accessed 7 Mar 2016.

Jaafari, Shirin. 2015. The Islamic State Needs Doctors and Engineers Too. Public Radio International. https://www.pri.org/stories/2015-05-21/islamic-state-needs-doctors-and-engineers-too.

Obama, Barack H. 2016. *Presidential Policy Directive—United States Cyber Incident Coordination*. https://www.whitehouse.gov/the-press-office/2016/07/26/presidential-policy-directive-united-states-cyber-incident. Accessed 1 Aug 2016.

Regional Focus. 2015. *Regional Focus Asia Pacific*. http://www.janes.com/article/39339/regional-focus-asia-pacific-es14e2. Accessed 18 Feb 2016; 17 June 2014.

Rose, M. 2015. Immunising the Mind: How Can Education Reform Contribute to Neutralizing Violent Extremism? British Council. https://www.britishco uncil.org/sites/default/files/immunising_the_mind_working_paper.pdf.

Rudner, Martin. 2013. Cyber-Threats to Critical National Infrastructure: An Intelligence Challenge. *International Journal of Intelligence and Counterintelligence* 26 (3): 453–481.

Shane III, Leo. 2016. Special Operations Command Eyeing Social Media as Next Battlefield. *Military Times*, March 2.

CHAPTER 11

Conclusion and Recommendations

Abstract This chapter explores the applicability of the Federal Secondary Data Case Study Triangulation Model and how well it worked regarding this research. Likewise, it goes into recommendations for future researchers to consider to build on this study.

Keywords Deep Fakes · Federal Secondary Data Case Study Triangulation Model · D.I.M.E.

The leadership of the U.S (and other countries aligned with it) must understand that the application of the D.I.M.E. is not a one-direction process. Those wanting to undermine and challenge the power and status of this country could equally use the instruments of national power against the United States to weaken its position; it is evident that the actors discussed throughout this book are leveraging cyber as an asymmetric tactic to do so. The implications for those that want to reduce the influence of the United States as a relevant leader in the world will most likely continue to exploit vulnerabilities in this nation's reliance on information technology (IT). In terms of defending against cyber threats, by assuming an adversarial role and through the use of the YIRTM-C (M), analysts and policymakers can 'red team' what potential belligerents can do to weaken a nation.

J. M. Weaver, *The U.S. Cybersecurity and Intelligence Analysis*, https://doi.org/10.1007/978-3-030-95841-1_11

Much of this country's critical infrastructure is comprised of systems and assets (both physical and virtual) that are so inextricably linked to the nation's national security that destruction of or an incapacitating attack on the nation would hinder its safety, public health, and economic well-being (GAO 2015; GAO-16-116T). Particularly vulnerable and exposed infrastructure include the financial and banking industry, energy sector, telecommunications networks, and more. This has opened a door for nefarious actors to prey on those systems that are reliant on IT, networks, the internet, and other computer technology. Current trends from this study show that state and non-state actors continue to probe for weaknesses in the public and private sector IT infrastructure. What is equally troubling is that the Defense Department has not historically developed military capabilities for the pointed mission of supporting civil authorities; accordingly, efforts to do so regarding cyber have been considered tepid and have lacked consistency with support provided for other exigencies that have mostly included logistics, engineering, medical and/or transportation (GAO-16–332 2016). These gaps could ultimately hinder defense support to civil authorities during times of emergency by adding to the time for the achievement of the restoration to normalcy.

The Director of Information Security Issues, Gregory Wilshusen, testified in October 2015 before the United States Congress on this country's electrical grid (GAO-16-174T 2015). When turning to a specific example, the potential adverse impact of cyber threats is exacerbated by the connections among industrial control systems, supervisory control and data acquisition (or SCADA) systems, the internet, information systems, as well as other infrastructure. This in turn creates opportunities for hackers to impede or disrupt critical services, which includes electrical power. The increased reliance on information technology systems and networks thereby exposes the weaknesses of the electric grid to potential cybersecurity vulnerabilities. Accordingly, cyber intelligence should be the primary purveyor of future cybersecurity missions (Mattern et al. 2014, 704).

The Federal Qualitative Secondary Data Case Study Triangulation Model was used to help the author strike a balance in his research approach when looking at the continuum of secondary data sources. As one can see from what is reflected in the chapter annexes, all components were considered and helped this author in answering the research questions.

That stated, the data reflected in Fig. 11.1 visually shows the breakout of the sources used in the analysis for each actor. The darker the color,

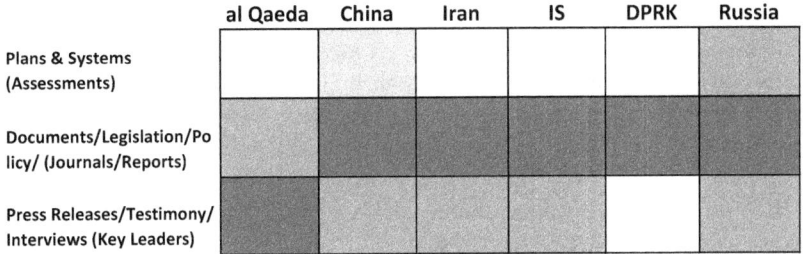

Fig. 11.1 Federal qualitative secondary data case study triangulation model matrix

the more secondary sources were available for the analysis that took place about that particular component of the Federal Qualitative Secondary Data Case Study Triangulation Model. Collectively and when looking at the three major components of this model, most of the evidence was derived from the documents, legislation, and policy area. Press releases, testimony, and interviews were used less but to a greater extent than the plans and systems component. Future researchers might want to focus more on plans and systems to see if anything changes.

There are several research implications that arise from this analysis. First, when turning to the YIRTM-C (M), most of the efforts looked to leverage information and military capabilities against the United States. Specifically, those that want to use asymmetry through cyber most likely see the relative inexpensiveness of information to hedge their strengths and promote one's causes (especially by non-state actors like IS). Likewise, many nation-states will either want to acquire their own military technology or leverage fissures in the interconnectedness and vulnerabilities of the U.S military systems. At present, China, Russia, Iran, and North Korea possess cyber weapons (GAO-19-204SP 2018, 9). One just needs to consider what could happen to the ability of a senior military commander to lead and control military operations if communication systems were turned off, if weapon systems couldn't communicate with one another, and/or if 'smart' bombs were rendered ineffective due to the lack of availability of global positioning satellites.

A second implication looks economically at how non-state and state actors are considering ways to hurt (or at least damage) the financial interests of the United States. One just needs to consider the attack on Sony Films in recent years and the intent of such actors as China, Russia,

North Korea, Iran, al Qaeda, and IS in acquiring information for their own economic expansion. Other trends look to the illegal acquisition of plans and systems of military equipment by hacking into defense contractors, and derivative cost savings particularly by state actors regarding their research and development efforts in their pursuit of cutting-edge weapon systems.

Diplomacy was an instrument lacking in the context of what is presently taking place in the world of cyber. Interestingly, Maurushat (2013, 122) points to this aspect and states that with all the talk about cyber 'war' and 'security' it is important to note that "there is little to nothing published comparatively on how cyber diplomacy or cyber peace might proceed." A reasonable interpretation could be that actors that are not as formidable as the U.S see a limited advantage to leveling the playing field and, therefore, will most likely want to use the benefit of cyber to their advantage.

Though qualitative studies look at answering questions like 'how' and 'why' and are not generalizable beyond the specific case, those that consider cyber as a potent threat might want to know more and could employ quantitative analysis in future to do so. An example could turn to China and Russia and the possible continuance of CNO by them against U.S equities for it (the United States) to spend more money on infrastructure and network defense to further weaken this nation economically. Secondly, China and Russia might also try to back door future weaknesses in C4ISR for others to exploit U.S vulnerabilities if hostilities were to break out (network vulnerabilities/back door intrusions, etc.). Moreover, they could exploit weaknesses to cripple financial sectors (though this might be unlikely due to the U.S. dollar being the primary reserve currency throughout most of the world and the inextricable linkages of globalization). Other efforts by Russia and China, in addition to those of other nation-states, may include the attempt to illegally acquire technology to reduce the high R&D costs associated with the development of military capabilities (like the J-20 stealth fighter in development by China as one example).

What could ensue in terms of defensive operations by the United States is the ongoing monitoring of threats, which will be necessary especially when considering the potential progression of cyber threats along what Mattern et al. (2014, 714) describe as level three (cyber incursion/surveillance), level four (cyber sabotage/espionage), and the potential escalation to level five (cyber conflict & war) operations. More-over, it appears that the U.S. DOD is taking a more offensive approach (CNA) to counter some non-state actors like IS as was covered in a speech given by a former U.S. Secretary of Defense when he essentially tasked the Cyber Command with its "first wartime mission" against the Islamic State (Carter2 2016).

What is necessary to defend against and effectively deter cyber threats? It becomes apparent that the private sector's reliance on networks (the nation's electrical power grid, hospitals, police stations, and more) as well as many non-defense & intelligence-related government agencies are seemingly vulnerable to attacks and need to build into their network greater resiliency. Though the U.S. Defense Department has been hacked in the past, the U.S. Defense Information Systems Agency (DISA) under the U.S. DOD has been effective at protecting the networks of this department and that of the president and vice president (for which DISA is also charged). On the civilian side, the National Cybersecurity and Communications Integration Center of the Cybersecurity and Infrastruc-ture Security Agency (CISA) under DHS, in the United States has a tall order as well to integrate and coordinate the whole-of-government approach for helping synchronize efforts and to mitigate effects from future attacks. Perhaps now is the right time elected leaders in the United States need to consider implementing a similar capability to protect all government networks and to study measures to more aptly guard from nefarious intrusions in the private sector as well. Cyber incident response is not enough; this nation must become even more proactive with allies to protect against attacks from the outset. Future research could also expand upon the results and findings within this book in other ways. Consider-ations can be afforded to studying the variables through the YIRTM-C (M) concerning other moderating variables to determine if similar results would be reached and can serve as confirmatory research. Moderating variables may include the following: legacy trade agreements, the sophis-tication of technology, treaties, et cetera; they might be relevant and have an impact on 'how' and 'why' cyber exploitation and attacks are used.

Moving forward, as a 'fifth domain,' cybersecurity safeguards must build upon three pillars: people, technology, and process, and these must be greatly expanded and improved upon for sophistication and network protection (Dutton 2017). Preventative technical measures could include frequently running antivirus software, active searches for identifying phishing emails, the use of anti-spyware on all computers, the installation of firewalls for internet connectivity, backing up critical databases, and securing Wi-Fi networks (Pomeroy 2019). Education of workers and the employment of more security-mindedness will reduce the risk of entities becoming victimized (Pomeroy 2019). The implementation of cross-domain checks like the Domain Name System (DNS) Risk Index that makes use of quantified metrics could help defense contractors identify gaps and improve resilience (Infoblox 2018). When used in conjunction with artificial intelligence (AI) and machine learning, tackling these challenges will not be insurmountable (Pomeroy 2019).

Cybersecurity vulnerabilities often are inherently technical but other issues are behavior-related (Dutton 2017). Such interconnectedness fosters a 'gray zone' that exists between the public–private partnership in terms of the allocation of responsibility and accountability of cybersecurity (Carr 2016). Moreover, it is incumbent for the United States to conduct timely and accurate risk analysis and arrive at sound assessments as this country looks to reduce the disturbances and disruptions of foreign cyber-attacks.

The information security solution must be a precursor of cybersecurity (Pomeroy 2019). Likewise, von Solms and Niekerk (2013) further separate cyberinfrastructure into a two-leveled structure and an ultimate security solution would involve the employment of information security at all levels and this is needed to add an extra layer which is a wider societal network for collective responsibility.

There are more implications for practitioners. As cyber becomes increasingly sophisticated, and through bots that are used to replicate messaging and 'deep fakes' that can look and sound exactly like real people (like world leaders who could profess false messaging or fake military orders to troops) separating fiction from reality will become even more problematic in the years to come. Ultimately a balance will need to be struck between 'convenience' and 'security' and the specifics identified in how to exactly strike that balance. Moreover, chief information officers (CIOs) and chief information security officers (CISOs) will shoulder the responsibility for helping protect their networks regardless of whether they are government, private, or other (Mattern et al. 2014).Cyber threats are not going away; they will most probably burgeon in the foreseeable future.

Recommendations (the Way Ahead)

There is an old expression that says once the genie has departed the bottle there is no way to put him back. Realistically, one cannot turn their back on technology; it is here to stay. The United States should look to build in redundancy by having backup systems that are not over-reliant on technology.

Kshetri (2013) writes that due to a decrease in conflict on the geopolitical stage, compounded with access to technology, one could suggest that digital and cyber-based threats will grow in both sophistication and prevalence to shape outcomes that could include diplomacy, military, and economic issues. That stated, things will most likely get worse before they become better. Issues emanating from disinformation and cyber-related threats will present several problems for the United States.

It is impossible to defend against every attack (Schwartz 2019, 69). Accordingly, measures could include bringing back into inventory older systems that are not linked and interconnected to offer levels of redundancy for vital infrastructure and systems. This is not to say that they should be used, but rather should be available for in-extremis scenarios. Threats can arise from nation-states and even non-state actors like terror and criminal organizations as they could use cyber-based tools to pursue the advancement of their positions.

What is more is that the United States should invest in training operators into how to roll out these systems (if required), ways in how to employ their capabilities, and in going beyond this—how to maintain them as well.

The United States should remain wary of diplomatic relations between actors covered in this book. Research has shown that Russia and China are increasing their cooperation between one another about trade, borders, and military sales (Røseth 2019, 268). Public and private sector organizations need to continue to have what is called continuity of operations plans (COOP) in place. This should include a layered approach such as backing up networks offline and even having hard copies of key information. This is particularly important to those levels of government and industry that work in the realms of national security and intelligence.

Over-reliance on cyber is a potentially grave issue. Compound this with the increases in the technical sophistication and complexity of cyber threats, and one can quickly see how the problem can magnify. Leaders, the legislative body, and the judicial system of the United States face a tall order to strike a balance between individual freedoms vs. state survival as this country looks to increase network resiliency within its borders. Cyber has policy implications at the domestic, bilateral, and international levels for all nations, and this includes the United States (Leuprecht et al. 2019, 382). To better prepare for cyber incidents, the government of the United States should look to (1) foster and maintain a culture of cybersecurity, (2) look to break down barriers to cybersecurity, (3) turn to and implement 'best practices' regarding cybersecurity, and finally (4) strive to find answers to the known unknowns (Norris et al. 2019). If the United States cannot outright eliminate the cliff, it should at least look to protect itself and minimize the fall from a CNA cataclysmic event.

References

Carr, Madeline. 2016. Public-Private Partnerships in National Cyber-security Strategies. *International Affairs* 92 (1): 43–62.

Carter2, Aston. 2016. Remarks on "Goldwater-Nichols at 30: An Agenda for Updating" (Center for Strategic and International Studies). http://www.defense.gov/News/Speeches/Speech-View/Article/713736/remarks-on-goldwater-nichols-at-30-an-agenda-for-updating-center-for-strategic. Accessed 11 Apr 2016.

DOS2. 2019. Cyber Issues. https://www.state.gov/policy-issues/cyber-issues/. Accessed 29 Dec 2019.

Dutton, Julia. 2017. Three Pillars of Cybersecurity. https://www.itgovernance.co.uk/blog/three-pillars-of-cyber-security. Accessed 31 Jan 2019.

GAO. 2015. Cybersecurity. http://www.gao.gov/key_issues/cybersecurity/issue_summary. Accessed 18 Feb 2016.

GAO-16–116T. 2016. Maritime Critical Infrastructure Protection DHS Needs to Enhance Efforts to Address Port Cybersecurity. http://www.gao.gov/assets/680/672973.pdf. Accessed 16 Feb 2016.

GAO-16–174T. 2015. Critical Infrastructure Protection Cybersecurity of the Nation's Electricity Grid Requires Continued Attention. http://www.gao.gov/assets/680/673245.pdf. Accessed 18 Feb 2016.

GAO-16–332. 2016. Civil Support DOD Needs to Clarify Its Roles and Responsibilities for Defense Support of Civil Authorities, during Cyber Incidents. http://www.gao.gov/assets/680/676322.pdf. Accessed 6 Apr 2016.

GAO-19–204SP. 2018. Report to Congressional Committees National Security Long-Range Emerging Threats Facing the United States as Identified by Federal Agencies. https://www.gao.gov/assets/700/695981.pdf. Accessed 23 July 2019.

Kshetri, N. 2013. Cybercrime and Cyber-security Issues Associated with China: Some Economic and Institutional Considerations. *Electronic Commerce Research* 13 (1): 41–49.

Leuprecht, Christian, Joseph Szeman, and David B. Skillicorn. 2019. The Damoclean Sword of Offensive Cyber: Policy Uncertainty and Collective Insecurity. *Contemporary Security Policy* 40 (3): 382–407.

Mattern, Troy, John Felker, Randy Borum, and George Bamford. 2014. Operational Levels of Cyber Intelligence. *International Journal of Intelligence and Counterintelligence*. 27 (4): 702–719.

Maurushat, A. (2013). From Cybercrime to Cyberwar: Security Through Obscurity or Security Through Absurdity? *Canadian Foreign Policy Journal* 19 (2): 119–122.

Norris, Donald F., Laura Mateczun, and Anupam Joshi. 2019. Cyberattacks at the Grass Roots: American Local Governments and the Need for High Levels of Cybersecurity. *Public Administration Review* 79 (6): 895–904.

Pomeroy, Jennifer Yongmei. 2019. Challenges of the U.S. National Security and Moving Forward (chapter). In *Global Intelligence Priorities (From the Perspective of the United States)*. New York, NY: Nova Science Publishers.

Røseth, Tom. 2019. Moscow's Response to a Rising China Russia's Partnership Policies in Its Military Relations with Beijing. *Problems of Post-Communism* 66 (4): 268–286.

Schwartz, Tamara B. 2019. A Dynamic Cyber-Based View of the Firm. Dissertation, Temple University.

von Solms, Rossouw, and Johan van Niekerk. 2013. From Information Security to Cybersecurity. *Computer and Security* 38: 97–102.

INDEX

© The Author(s), under exclusive license to Springer Nature Switzerland AG 2022
J. M. Weaver, *The U.S. Cybersecurity and Intelligence Analysis*, https://doi.org/10.1007/978-3-030-95841-1

CPSIA information can be obtained
at www.ICGtesting.com
Printed in the USA
LVHW080852070322
712805LV00004B/296

9 783030 958404